Ontology in the AEC Industry

A Decade of Research and Development in Architecture, Engineering, and Construction

Sponsored by
Technical Council on Computing and Information Technology
of the American Society of Civil Engineers

Edited by
Raja R. A. Issa, Ph.D., J.D., P.E.
Ivan Mutis, Ph.D.

ASCE

Published by the American Society of Civil Engineers

Library of Congress Cataloging-in-Publication Data

Ontology in the AEC industry : a decade of research and development in architecture, engineering, and construction / sponsored by Technical Council on Computing and Information Technology of the American Society of Civil Engineers ; edited by Raja R. A. Issa, Ph.D., J.D., P.E. Ivan Mutis, Ph.D.
pages cm
Includes bibliographical references and index.
ISBN 978-0-7844-1390-6 (pbk.) — ISBN 978-0-7844-7912-4 (pdf)
1. Communication in engineering. 2. Civil engineering. 3. Ontology. I. Issa, Raymond. II. Mutis, Ivan.
TA158.5.O58 2015
601′.4—dc23

2015009320

Published by American Society of Civil Engineers
1801 Alexander Bell Drive
Reston, Virginia, 20191-4382
www.asce.org/bookstore | ascelibrary.org

Errata: Errata, if any, can be found at http://dx.doi.org/10.1061/9780784413906.

ISBN 978-0-7844-1390-6 (print)
ISBN 978-0-7844-7912-4 (PDF)
Manufactured in the United States of America.

19 18 17 16 15 1 2 3 4 5

Contents

Preface v

Management (Integration, Sharing, Exchanging) of Semantically-Expressive Information

1. **Domain Knowledge–Based Information Retrieval for Engineering Technical Documents 1**
S.-H. Hsieh, K.-Y. Lin, N.-W. Chi and H.-T. Lin

2. **Wiki-Con: A Pragmatic Approach for Semantic Interoperability for Construction Projects 27**
I. Mutis and R.R.A. Issa

3. **Ontology to Support Healthcare Facility Management.....................47**
J. Lucas and T. Bulbul

4. **Ontology to Support Multi-Objective Integrated Analyses for Sustainable Construction: A Conceptual Framework 73**
Y. Zhu

Theoretical Propositions in Ontology Research

5. **From Deep Blue to Watson: The Nature and Role of Semantic Systems in Civil Informatics 97**
T. E. El-Diraby

Innovative Approaches to Relate Context to Ontology

6. **The Potential of Ontology-based Serious Game Design for the AEC Domain 121**
K. Schatz and U. Rüppel

Issues on Ontology Development and Validation

7. **Fieldwork-Based Method for End-User Engagement in Domain Ontology Development... 149**
L.L. olde Scholtenhuis and T. Hartmann

Ontology Research to Enable Collaboration

8. **Transaction Formalization in Infrastructure Management using an Ontological Approach... 169**
J. Zeb and T. Froese

9. A Framework for Regulatory Ontology Construction within AEC Domain 193
L. J. McGibbney and B. Kumar

10. Taxonomy Development toward the Domain Ontology of Construction Contracts: A Case Study on AIA A201-2007 217
J. Niu, R.R.A. Issa and I. Mutis

Index 251

Preface

Civil and construction engineering firms have relied on information technology to efficiently represent and manage information (e.g., sharing, exchanging, and integrating information). Information technologies support the representation of information and information management activities (e.g., retrieving, storing, reproducing, and exchanging). The basic assumption is that the use of technologies enables project stakeholders to seamlessly comprehend and manipulate information with minimal inconsistencies, ambiguities, and delays and disruptions, by incorporating richness of detail for particular applications to enable more informed decisions. However, technologies poorly capture and represent the semantics of shared information to cope, for instance, with heterogeneous and poorly structured information.

Ontologies formalize and represent information that is valuable to the end-users. Ontologies formalize concepts by conforming to a common vocabulary or a set of rules. They allow participants to map the domain concepts to a computable format and to base their business information on the formalization of concepts as a reference. However, the ontology construction and implementation process is challenged by significant problems. The architecture, engineering and construction (AEC) community, therefore, requires further research efforts to address the challenges by proposing more adequate approaches to efficiently create, share, integrate, and reuse knowledge in the AEC domain.

For example, the formalization and generation of information content have the same patterns and relationships for one particular community, as they evolve in local forms of vocabulary with a set of boundaries (contexts) that define the terms, but others might not have the same perceptions of the patterns and relationships. These basic-form differences lead to problems of reconciliation of the concepts into an ontology-standard (multiple competing ontologies, dynamic expansion of the ontologies, time-consuming consensus reaching process, e.g., quality of community generated content), and also other barriers to the representation of expressiveness and reasoning. The use of ontology as a shared conceptualization has the potential to be a solution to represent and manage information. However, the aforementioned limitations significantly hinder the construction of well-founded and tractable ontologies to fully deploy ontology uses at different levels of awareness (including the users' perception and comprehension) and in a more dynamic, complex real world contexts such as are the civil infrastructure and construction projects.

This publication aims to offer a comprehensive overview of the recent advances in ontology research in the AEC domain that have focused on enabling information exchange, sharing, and integration, and on facilitating users' understanding of the semantics of information in any project activity. Towards this end, the included chapters on relevant ontology research focus on new ideas, including theoretical and visionary propositions, into the management (integrating, sharing, and exchanging) of semantically-expressive information, foundational-driven practical ontologies, relating context to ontology, evaluation methods and metrics of AEC domain ontologies and issues of their validation and collaboration.

R.R.A. Issa
I. Mutis

CHAPTER 1

Domain Knowledge–Based Information Retrieval for Engineering Technical Documents

Shang-Hsien Hsieh*
Ken-Yu Lin†
Nai-Wen Chi*
Hsien-Tang Lin*

Abstract: *Technical documents with complicated structures are often produced in architecture/engineering/construction (AEC) projects and research. Information retrieval (IR) techniques provide a possible solution for managing the ever-growing volume and contexts of the knowledge embedded in these technical documents. However, applying a general-purpose search engine to a domain-specific technical document collection often produces unsatisfactory results. To address this problem, we research the development of a novel IR system based on passage retrieval techniques. The system employs domain knowledge to assist passage partitioning and supports an interactive concept-based expanded IR for technical documents in an engineering field. The engineering domain selected in this case is earthquake engineering, although the technologies developed and employed by the system should be generally applicable to many other engineering domains that use technical documents with similar characteristics.*

We carry out the research in a three-step process. In the first step, since the final output of this research is an IR system, as a prerequisite, we created a reference collection which includes 111 earthquake engineering technical documents from Taiwan's National Center for Research on Earthquake Engineering. With this collection, the effectiveness of the IR system can be further evaluated once

---

*Dept. of Civil Engineering, National Taiwan University, No. 1, Roosevelt Rd., Sec. 4, Taipei City, 10617, Taiwan; Tel: +886-2-3366-4313; Fax: +886-2-2739-0534; Email: shhsieh@ntu.edu.tw, naiwenchi@ntu.edu.tw, d93521003@ntu.edu.tw

†Dept. of Construction Management, University of Washington, 120 Architecture Hall, Box 351610, Seattle, WA 98195-1610, USA; Tel: +1-206-616-1915; Fax: +1-206-685-1976; Email: kenyulin@uw.edu

it is developed. In the second step, the research focuses on creating a base domain ontology using an earthquake-engineering handbook to represent the domain knowledge and to support the target IR system with the knowledge. In step three, the research focuses on the semantic querying and retrieval mechanisms and develops the OntoPassage approach to help with the mechanisms. The Onto-Passage approach partitions a document into smaller passages, each with around 300 terms, according to the main concepts in the document. This approach is then used to implement the target domain knowledge–based IR system that allows users to interact with the system and perform concept-based query expansions. The results show that the proposed domain knowledge–based IR system can achieve not only an effective IR but also inform search engine users with a clear knowledge representation.

INTRODUCTION

Technical documents in specific domains, such as journal papers, patents, technical reports, or domain handbooks, are produced during architecture/engineering/construction (AEC) projects and research. To make knowledge management of these more efficient, search engines are often applied to retrieve needed documents. Information Retrieval (IR) is the process to retrieve specific information according to a given information need. The most popular application is text-based IR. In most cases, search engine users also present their information request through query terms written in plain text.

Although most people have used general-purpose search engines, such as Google, Yahoo!, or Bing, applying a general-purpose search engine to a domain-specific document collection often produces unsatisfactory results. Technical documents, for example, have complicated structures and multiple concepts; these are the two most important causes of low IR effectiveness. Callan (1994) defined such documents with complicated structures and multiple concepts as "Large Documents". When a search engine system is used to search for such large documents, it often leads to two problems. The first is that multiple topics in a document lower the document's relevance ranking if a search engine user requests or cares about only one query topic. Moreover, if the search engine user sets a higher threshold to filter the retrieved documents, such a document may be neglected because its rank has been lowered. The second problem is that even if a relevant document is found, the unclear presentation of the search results may trouble search engine users as well. It is not easy for search engine users to locate the relevant paragraphs because they might be scattered over the entire document (Callan 1994). In this situation, the IR task cannot produce satisfactory results because it cannot precisely retrieve the information. Partitioning a long document or a document with multi-topics into smaller passages as the base unit for search engines helps retrieve target information more precisely and efficiently. A special-purpose search engine supported by domain knowledge is a possible solution to implement the document partitioning mechanism.

In this chapter, we use earthquake engineering as an example to demonstrate the construction of a special-purpose IR system. The system consists of three important steps: (1) the reference collection that evaluate the search engine; (2) the domain knowledge used to support the IR system; and (3) the search engine system that implements the algorithm to effectively use the domain knowledge. In the following sections, this paper will introduce the three main research pillars. The first is the reference document collection, which consists of 111 earthquake engineering documents. In order to evaluate the effectiveness of the IR system, such a document collection should be labeled by domain experts and constructed as a "reference collection". Domain experts must perform several relevance assessments to build the relationships between each document and each information request so that one can tell whether the IR system retrieves the right documents. The second step is to construct the domain knowledge behind the IR system. We use the *Earthquake Engineering Handbook* (Chen and Scawthorn 2003) to demonstrate the semi-automated knowledge acquisition process. In the third step, based on the domain knowledge extracted from the domain handbook, we illustrate how to apply the domain knowledge to implement the IR system, with a view to addressing the problematic complex structure of technical documents. The results indicate that the proposed domain knowledge–based IR system can achieve not only effective IR but also a clear knowledge representation for search engine users.

THE NCREE REFERENCE COLLECTION

In this research, the authors develop a domain knowledge–based IR system for technical documents. Before developing the IR system, a reference collection must be built up for evaluating the IR effectiveness. After considering the available resources, we chose an existing document database to create the reference collection. Taiwan's National Center for Research on Earthquake Engineering (NCREE) provided 111 technical reports for the NCREE reference collection (see Table 1-1).

To further evaluate the IR effectiveness of the mentioned search engine system, the NCREE reference collection needs three important components: a set of documents, a set of information requests, and answers to information requests. The purpose of an information request is to reproduce an actual information need (Mizzaro 1998), and it can be expanded into various query terms for search engines. The set of documents relevant to an information request is the answer to that request, which is used to determine the IR effectiveness. Therefore, it can be treated as a criterion to evaluate whether a search task is successful and the corresponding precision and recall indices can be calculated. The following section will illustrate how we define the information requests and how we conduct

Table 1-1. The properties of the NCREE reference collection (Adapted from Lin et al. 2012, 2008)

Property	*Abstract*	*Full-text*
Collection Size	720 KB	6.25 MB
Language	Chinese & English	
Num. of Documents	111	
Num. of Information Requests	8	
Avg. Num. of Eng. Terms per Document	148	1478
Avg. Num. of Eng. Terms per Info. Request	125	
Num. of Unique Eng. Terms in the Collection	3158	20990

relevance assessments between the documents and requests in order to develop the NCREE reference collection into a reference collection.

DEFINING INFORMATION REQUESTS

To define information requests for the NCREE reference collection, a group of domain experts manually assigned a number of concepts in the specified area of interest (*i.e.*, earthquake engineering) to each document in the collection. The experts then aggregated closely related concepts to form candidate information requests. Any candidate information request with too few or too many associated documents was discarded and the remaining requests were subsequently refined by the experts. Once the completed relevance assessment of a candidate information request passed an inter-judge consistency test, the request was officially accepted. The inter-judge consistency test is explained later on under assessing document relevancies.

When the NCREE reference collection was compiled, there were sixteen candidate information requests at the beginning of the process, and five of which were later discarded for having too few associated documents. For each information request, four metadata were added to the information request to make it clearer. The first was "topic", which defined the topic that a request involves. The second was "description", which provided a short description as the definition of the topic. The third was "narrative", which gave more details on the criteria for judging document-topic relevance. The last was "concepts", which listed several related domain concepts for the information request. A sample information request is shown in Table 1-2.

Assessing Document Relevancies

After defining the information requests for the NCREE reference collection, the next step was to conduct relevance assessments between each document and each information request. When preparing the NCREE reference collection, we invited a panel of fifteen experts from the NCREE to participate in the relevance

Table 1-2. A sample information request in the NCREE reference collection

No.	3
Topic	Seismic Experiment
Description	A relevant document will mention experiments concerning bridges, buildings, other types of structures, or non-structural components against artificial seismic ground motions. The structures or non-structural components may be built full-scale in such an experiment.
Narrative	Often a Shaking Table Test is employed in the relevant documents to simulate a specific earthquake as a seismic wave. The wave may be inputted into the test target, such as a structure or a non-structural component, to investigate the seismic behaviors of the test target or its components.
Concepts	Shake Table Test, Shaking Table Test, Pseudo-dynamic Experiment, Pseudo-dynamic Test.

assessment workshop. Depending on the expert's background and research interests, each expert performed a relevance assessment for at least one, and at most three, information requests during the one-hour workshop. In addition, at least two experts, including the original drafter of the information request, evaluated each information request. The aim of these experts was to make a binary decision to determine whether each document in the collection was relevant to the candidate information request. The project team adopted a binary judgment during collection preparation because this is the fundamental assumption of many retrieval performance indices.

To synchronize the definition of document relevance among the participating experts, the workshop included a short session that introduced an example scenario concerning the relevance judgment. The scenario asked the experts to picture a situation where they had to write a report on the subject of the given candidate information request. If a document contained information that they would use to prepare the report, then the document was considered relevant. After the workshop, the overlap value (Voorhees 1998) and Cohen's Kappa measure (Sheskin 2003) were calculated to verify the experts' assessments (Lin, et al. 2008). Cohen's Kappa measure shows internal consistence between relevance assessments. Several information requests were removed because their relevance assessments were dramatically different among the participating domain experts. In the end, eight out of eleven information requests were deemed satisfactory.

Discussions

This section introduces the importance of the reference collection. Although documents for a specific domain are available, a reference collection with the assessments of domain experts must first be developed in order to evaluate the IR effectiveness. The intention is to support AEC text-based IR developments by initiating research on the creation of domain-specific reference collections.

For the NCREE reference collection, because the volume of source documents in the collection increases annually, we will incrementally expand the created collection by including newly published NCREE technical reports. However, we will also need to update the relevance assessment results when adding new documents to the collection. Depending on the available resources, we might invite the domain experts to update the assessments for every document in the collection or simply conduct the assessments on the newly added collection documents themselves. In addition, new information requests can be introduced. In that case, a relevance assessment for a new information request will be generated from scratch.

In this chapter, the NCREE reference collection is not only applied to evaluate the domain knowledge-based search engine but also serves as the document collection that the earthquake engineering domain users can search against. And since these documents belong to the earthquake engineering domain, the domain knowledge must be organized to support the search engine system. In the next section, the authors will introduce how to develop an earthquake engineering domain ontology to represent the domain knowledge. With the domain specific document collection (i.e. the NCREE reference collection) and the domain knowledge (i.e. the earthquake domain ontology discussed in the following section), the domain knowledge-based search engine can be implemented as illustrated under the Ontopassage section.

THE BASE DOMAIN ONTOLOGY FOR EARTHQUAKE ENGINEERING

Step two of our research was to construct the domain knowledge to assist the special-purpose search engine. We chose an ontology as the knowledge representation. Gruber (1993) defined ontology as "a formal, explicit specification of conceptualization". The "conceptualization" described in this definition refers, in simple and logical terms, to the overarching worldly domain. Issa and Mutis (2006) explained, "Conceptualizations are described by a set of informal rules used to express the intended meaning through a set of domain relations." Many domains have tailored domain ontologies to suit their specific requirements. In the civil engineering domain, the e-COGNOS project successfully presented the ontology application of knowledge management and information retrieval (Lima, et al. 2005). Several construction domain ontologies have also been developed for various practical purposes, such as knowledge management (Diraby, et al. 2005; El-Diraby and Kashif 2005), interoperability among different software systems (*e.g.*, CAD and GIS systems) (Peachavanish, et al. 2006), and project collaboration (Garcia, et al. 2004). Outside the engineering domain, using ontology to assist with information retrieval tasks through for example the annotation of semantic information (Gruber 1995) has been widely explored.

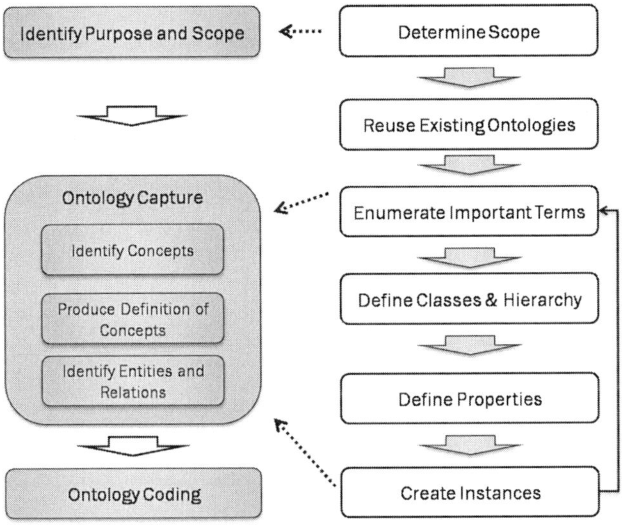

Figure 1-1. General ontology development procedures by (left) Uschold and Gruninger (1996) and (right) Noy et al. *(Noy, et al. 2001; Hsieh, et al. 2011)*

Figure 1-1 shows a general procedure for developing a domain ontology summarized from two well-known exemplars. The left side of Figure 1-1 was proposed by Uschold and Gruninger (1996), and the right side was suggested by Noy *et al.* (2001). Both procedures imply that the scope of the domain knowledge should be defined first. The major objective of developing an ontology is to define concepts/class, entities/instances, and relations/hierarchy in a knowledge domain.

The participation of domain experts is also crucial in the procedure for developing domain ontologies. The iterative procedure, from enumerating important terms and defining concepts, to identifying hierarchies and relationships, requires an intense effort by the participating experts with respect to the discussion and subsequent formulation of revisions. Mindful of this fact, we were motivated to create an approach to ontology development that could reduce the burden on the participating experts.

Although an increasing number of civil and infrastructure engineering specifications or information standards have been established to describe the structure components and construction contracts, they also serve as useful resources for ontology development by providing detailed domain knowledge. Recognizing the potential of these resources and the limited amount of research in the civil engineering domain concerned with ontology creation, we propose an ontology development solution using domain handbooks.

These handbooks, replete with domain knowledge, were of course edited by experienced domain experts. Each volume not only records the domain knowledge but also provides more professional content such as terminologies and definitions that are related to a domain than just the generic web pages. This observation inspired us to devise a rapid, straightforward procedure for constructing the base domain ontology by reusing the domain handbooks as knowledge resources. An immediate challenge for the proposed procedure was how to competently handle the large number of text documents featured in a handbook. Digitizing the complete contents of domain handbooks by OCR or typing in is too laborious and is unfeasible if a digital version of the handbook cannot be acquired. It was decided that the process could be expedited if an ontology could be generated from only the table of contents, definitions, and the index of a domain handbook. The approach is innovative in that it not only takes advantage of existing resources (*i.e.*, extracting knowledge from domain handbooks), but also reduces the workload of the participating experts, who are often consulted to define the complex relationships between concepts and instances during ontology development.

We used the *Earthquake Engineering Handbook* (Chen and Scawthorn 2003) as an example in order to propose a practical approach from the engineers' point of view. The objective here was to establish a general draft (or base) of a domain ontology from an engineering domain handbook, which defines the scope and knowledge representation space of the targeted domain. The proposed procedure is intended to extract important terms or phrases and build relationships between them, and then organize the phrases and the relationships into a phrase map before expert assessment and evaluation. These terms came from only the table of contents, definitions, and the index of a domain handbook, and are manually gathered through document scanning and OCR. After the manual procedure, a semi-automated filtering mechanism which contains three weighting rules is then applied to distinguish the properties of the terms (i.e., whether the term plays the role of concept or instance in the ontology) and build up the relationships between these terms. After that, we stored the terms, relationships, and hierarchy information as a phrase map in the XML format. These relationships and hierarchies could thereby be represented as a tree or net graph with the XML document, which could then be easily transformed into the final base domain ontology in a web ontology language (OWL) format, OWL Lite. The phrase map was then evaluated by domain experts and revised as the base domain ontology.

An evaluation of the phrase map by the domain experts is also considered an effective mechanism for obtaining expert feedback. Experts from NCREE were invited to participate in the evaluation phase of the base domain ontology development procedure. Such an approach is also an iterative procedure because the evaluation results could potentially provide feedback to assist with the revision of the base domain ontology.

The Approach

The main objective of the present approach is to produce a good base domain ontology that, in practical use, can reduce the workload of ontology development for participating experts. Figure 1-2 illustrates the overall semi-automatic procedure for constructing a base domain ontology from domain handbooks, and Table 1-3 summarizes the main task of each stage. A domain handbook

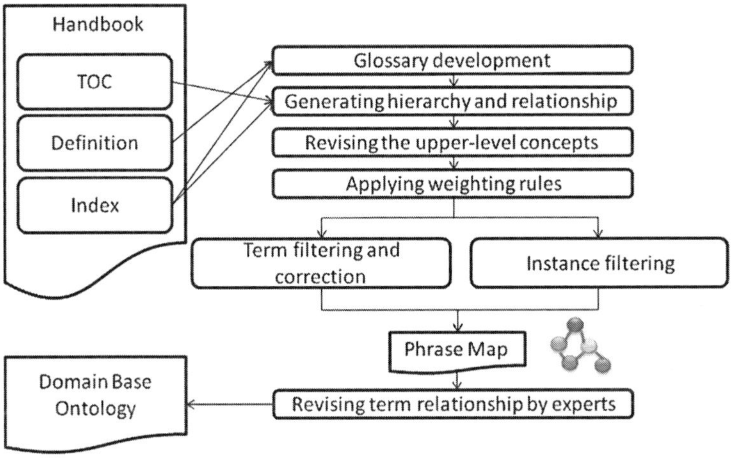

Figure 1-2. Overall procedures for developing a base domain ontology from handbooks (Hsieh et al. 2011)

consisting of the "Table of Contents" (TOC), "Definition" or description of domain-specific terms, and the "Index" of these terms is digitized (if the digital version of the handbook is unavailable, we scan the essential parts of the handbook and then use OCR to acquire the digitalized text) to provide the initial input for the ontology development. The proposed procedure is intended to extract important terms or phrases and build the complex relationships between them before expert assessment and evaluation. The details of the five important stages in Figure 1-2 will be further discussed in the following subsections.

Glossary Development

To facilitate computational data processing, as indicated by Step 1 in Figure 1-2, all the necessary information was gathered and digitized. The handbook consisted of five technical sections, thirty-four chapters, and 1,512 pages. Each chapter was consigned to one of the five sections, including "Fundamentals", "Geoscience Aspects", "Structural Aspects", "Infrastructure Aspects", and "Special Topics". Figure 1-3 illustrates the upper-level concepts extracted from the TOC.

Furthermore, each chapter was supplemented with a glossary that contained 2 to 20 domain-specific term definitions (See Figure 1-4a). Domain related phrases, such as people, places, events and technologies, could then be extracted from these term definitions. In addition, the handbook's Index section also provided rich domain related phrases (See Figure 1-4b).

After the *Earthquake Engineering Handbook* was completely processed, 2,880 phrases were collected, 679 phrases were drawn from term definitions, and 2,201 phrases were taken from the handbook index. The authors finally store these terms in an XML file as shown in Figure 1-4c.

Table 1-3. Summary of the semi-automated procedure for developing a base domain ontology from handbooks (Hsieh et al. 2011)

Stage	*Type*	*Description*
Glossary Development	Semi-automated	The terms are first gathered and digitized (by scanning and OCR) from a handbook according to the structure of the TOC.
Generating Hierarchy and Relationships	Automatic	Hierarchical relationships of terms are initially established.
Revising the Upper-Level Concepts	Manual	Experts review the upper-level concepts and relationships between the concepts extracted from the handbook.
Applying Weighting Rules, *e.g.*, Filtering and Term Filtering	Semi-automated	Weighting rules are applied to automatically determine whether key terms extracted from the handbook are concepts or instances, and domain experts review the results.
Evaluation and Refinement	Manual	Experts review the base ontology and refine the relationships between the concepts.

Generating Hierarchies and Relationships

In the second step, the relationships between the phrases of the domain glossary furnish a domain vocabulary, while acting as a form of knowledge representation. The glossary collected from the first step contained phrases featured in either the chapters or the index. Any attempt to judge the exact relationship between two phrases was an inherently difficult and complicated enterprise. To alleviate this problem with respect to the glossary phases, we strategically generalized these relationships into two types: the "is-a" and the "has-a" relations. The benefit of such generalizations was that term relationships could be quickly picked up from the handbook to highlight the hierarchy between terms. In this stage, the authors only focused on storing the complicated relationships for further analysis and operation. For instance, our generalization strategy dictates that if A is a bridge and B is a pier, the relationship B is a "part-of" A (i.e., 'a pier is one part of a bridge') can be simplified into "A 'has-a' B". Leveraging the concept of relationship

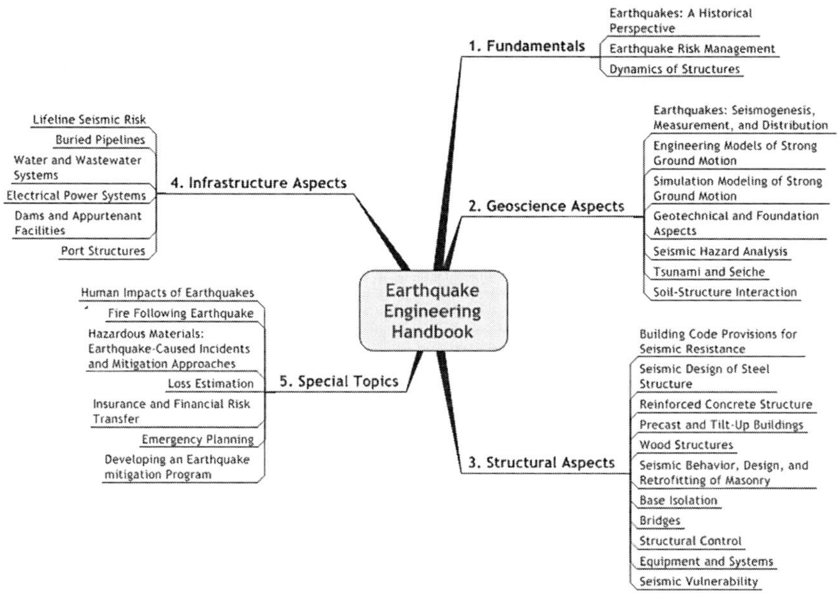

Figure 1-3. Concept map extracted from the TOC of the Earthquake Engineering Handbook *(Hsieh, et al. 2011)*

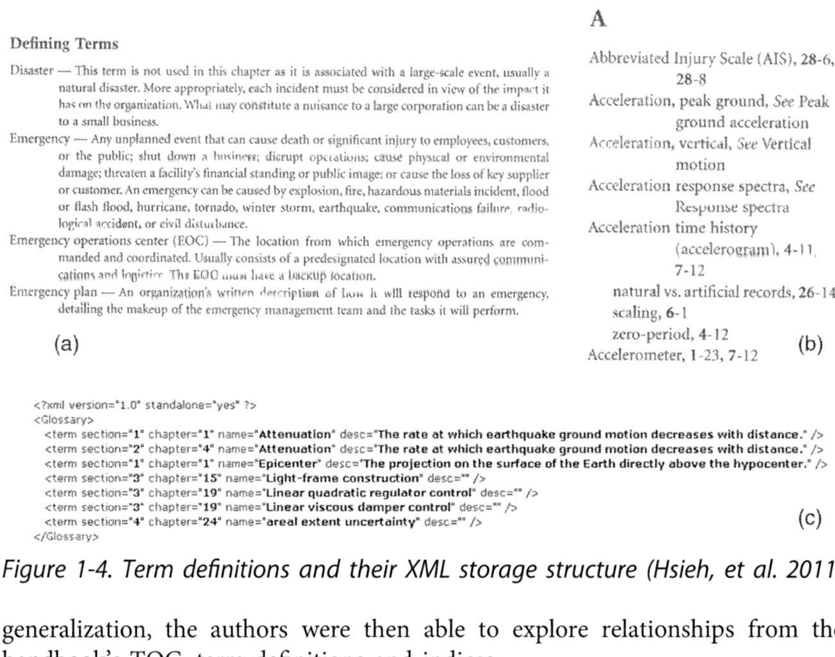

Defining Terms

Disaster — This term is not used in this chapter as it is associated with a large-scale event, usually a natural disaster. More appropriately, each incident must be considered in view of the impact it has on the organization. What may constitute a nuisance to a large corporation can be a disaster to a small business.

Emergency — Any unplanned event that can cause death or significant injury to employees, customers, or the public; shut down a business; disrupt operations; cause physical or environmental damage; threaten a facility's financial standing or public image; or cause the loss of key supplier or customer. An emergency can be caused by explosion, fire, hazardous materials incident, flood or flash flood, hurricane, tornado, winter storm, earthquake, communications failure, radiological accident, or civil disturbance.

Emergency operations center (EOC) — The location from which emergency operations are commanded and coordinated. Usually consists of a predesignated location with assured communications and logistics. The EOC must have a backup location.

Emergency plan — An organization's written description of how it will respond to an emergency, detailing the makeup of the emergency management team and the tasks it will perform.

(a)

A

Abbreviated Injury Scale (AIS), 28-6, 28-8
Acceleration, peak ground, *See* Peak ground acceleration
Acceleration, vertical, *See* Vertical motion
Acceleration response spectra, *See* Response spectra
Acceleration time history (accelerogram), 4-11, 7-12
natural vs. artificial records, 26-14
scaling, 6-1
zero-period, 4-12
Accelerometer, 1-23, 7-12

(b)

```
<?xml version="1.0" standalone="yes" ?>
<Glossary>
  <term section="1" chapter="1" name="Attenuation" desc="The rate at which earthquake ground motion decreases with distance." />
  <term section="2" chapter="4" name="Attenuation" desc="The rate at which earthquake ground motion decreases with distance." />
  <term section="1" chapter="1" name="Epicenter" desc="The projection on the surface of the Earth directly above the hypocenter." />
  <term section="3" chapter="15" name="Light-frame construction" desc="" />
  <term section="3" chapter="19" name="Linear quadratic regulator control" desc="" />
  <term section="3" chapter="19" name="Linear viscous damper control" desc="" />
  <term section="4" chapter="24" name="areal extent uncertainty" desc="" />
</Glossary>
```

(c)

Figure 1-4. Term definitions and their XML storage structure (Hsieh, et al. 2011)

generalization, the authors were then able to explore relationships from the handbook's TOC, term definitions and indices.

After extracting and accumulating all three relationship components (i.e., the two terms and the relationship between the two terms), the terms and relationships could be presented in combination as a complex hierarchical concept network. By the end of this step, 4,508 relationships had been generated from the handbook.

Revising the Upper-Level Concepts

To make the developed earthquake engineering ontology more complete and accurate, seventeen experts from the National Taiwan University and the NCREE were invited to revise the upper-level concept map based on the extracted TOC of the *Earthquake Engineering Handbook*. During the review and revision process, there were three recommended principles: (a) each term or phrase shown in the upper-level concepts should be an independent concept; (b) the chosen terms or phrases should be frequently used by the developed glossary; and (c) the relationships between concepts should not be limited to one-to-one. The experts reorganized the first-level concepts as "Geoscience", "Geotechnical Earthquake Engineering", "Seismic Technologies", "Structure", and "Earthquake Risk Management". In comparison with the TOC hierarchy, the "Fundamentals" was renamed as "Geoscience", "Infrastructure Aspects" was integrated within "Structure", and some technical issues were combined into "Seismic Technologies". Regarding hazards, emergency planning and loss assessment issues that appeared in "Special Topics", the experts arranged these topics as "Earthquake Risk Management". Two or three experts then further reviewed each descendant concept of the first-level concepts in accordance with their professional specialties. We then merged the upper-level concepts with term relationships. This ensured greater clarity and consistency for the earthquake engineering ontology that was produced.

Applying Weighting Rules for Instance and Term Filtering

Prior to exhaustive revision by our panel of experts of the terms and their relationships, a filtering process was applied to reduce the number of term duplicates. These duplicates were the results of different forms of a single word (*i.e.*, terms appearing in both singular and plural forms and different verb tenses) and can be removed by stemming. While developing the glossary in the first step, we preliminarily eliminated the duplicated terms in the collected data. We then identified and clustered the terms with similar concepts to help examine and eliminate the redundant concepts in the developed phrase map.

In ontology development, concept and instance are two important elements. They often form a "is-a" relationship. For example, from the sentence "London Bridge is a bridge", we can distinguish that "bridge" is a concept while "London bridge" is an instance. Determining whether a term or phrase was a concept or an instance posed further difficulties. Because almost all terms had direct relationships with all thirty-four chapter titles, there was a risk of unmanageability if each title became a huge cluster with too many child nodes. These huge clusters therefore needed to be partitioned into more hierarchical levels to present a more logical distribution of concepts. This research designed an evaluation and refinement procedure to: (a) filter out extracted terms that might be produced by typing errors; (b) integrate synonymous terms and phrases into a single concept; (c) differentiate concepts and instances; and (d) facilitate the experts' task of defining classes, properties, instances, and their exact relationships, subject to appropriate revision. The phrase map, consisting of 2,547 terms and 4,508 relationships, served as input for the procedure. We implemented three weighting rules to score each term or phrase for term filtering, the generation of concepts, and the filtering of instances (Hsieh, et al. 2011). In order to illustrate the weighting rules in this chapter, the phrase network is simplified as a tree model as shown in Figure 1-5.

In the first weighting rule (i.e., Figure 1-5a), the number of relationships each term had was assigned to the term as its base score. Therefore, a term's score was solely dependent on the number of relationships. With this simple rule, the erroneous operation (e.g., typos of terms or phrases, and missing connections of relationships) could be found by examining nodes that scored 0 points. If a node scored only 1 point (e.g., nodes I, H, J, F), it had to be located at the end of the tree, meaning that the term or phrase that the node represented was probably an instance. On the contrast, if a node got a higher score (e.g., nodes C and D), it implied that this term or phrase could be more general, with a potential to be a concept.

Considering that the direction of a relationship could also influence the hierarchical level of the connected nodes, the authors applied the second

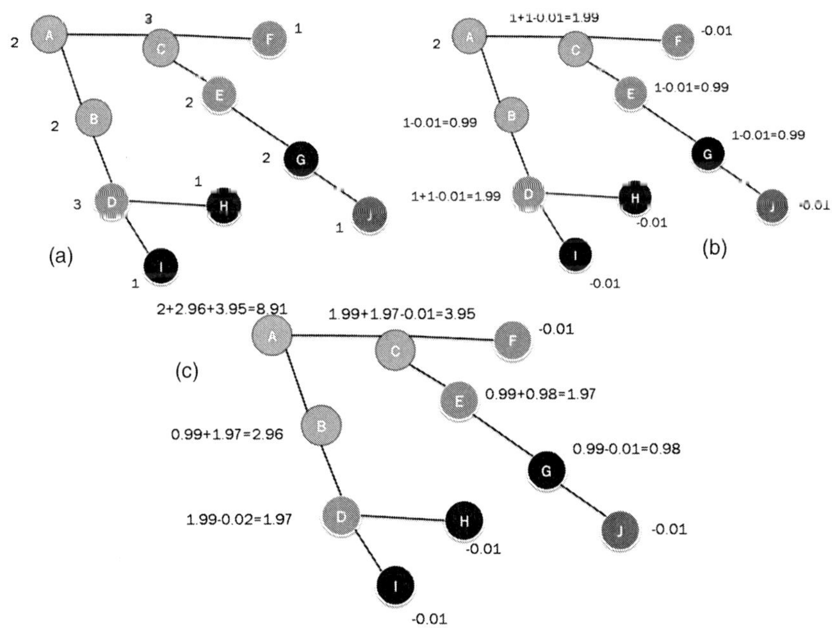

Figure 1-5. Three term weighting rules for ontology evaluation (Hsieh et al. 2011)

weighting rule to assign positive or negative scores to a node, depending on the type of relationships the node was attached to. As shown in Figure 1-5b, a node received a positive score if it was attached with a "has-a" relationship; if an "is-a" relationship was connected to the node, it received a negative score. In the example shown in Figure 1-5b, for illustrative purposes, the authors arbitrarily selected 1 and −0.01 as the positive and negative scores. In keeping with the second weighting rule, we could easily know how many "has-a" and "is-a" relationships are connected to a term. The third weighting rule enacts a propagation of the positive/negative scores assigned by the second rule (see Figure 1-5c). As stipulated by the third rule, the score of an upper-level node was the sum of its own score and all scores of its lower-level child nodes. Evaluating the entire phrase map using the three rules helps clarify if a term or phrase was essentially a concept or an instance. For example, in Figure 1-5c, even though nodes D, E and F were all at the third-tier of the tree, node F obviously could have a much higher chance to be an instance. Meanwhile, although nodes B and C were both at the second-tier of the tree, the fact that node C received a higher score because the child nodes presented a deeper (or larger) sub-tree, implied that either: (a) it was more adequate to select node C as a concept, or (b) node C probably had an improper hierarchy that needed to be revised.

After applying the three weighting rules, twenty-four terms were found to be term duplicates and were therefore removed from the glossary. Of the remaining 2,523 terms, 2,025 were recognized as probable instances and 498 were suggested as concepts.

Revising Term Relationships by Domain Experts

In the expert assessment step, we invited seventeen experts from the National Taiwan University and the NCREE to a workshop to revise the extracted terms and relationships. The authors selected the nodes that received the top-20 score as assessment categories. Each category containing 83 terms on average was assessed by one or two experts according to their research interests and background. Experts were asked to score each relationship (i.e., the relationship between each term and category) from 1 to 3 according to the intensity of the terms' correlation. For each term and phrase, experts also provided suggestions about the wording, synonyms and recommended classification. After the assessment workshop, we gathered the scoring data and suggestions, and calculated the scores of the categories assessed by the experts. Relationships receiving a 0 score were removed from the generated hierarchy. The assessment results were then used to refine the weighting of the third rule for evaluating the relationships. The reduction factors 0.33, 0.66 and 1 were multiplied by the scores of the original relationships after the relationships were scored 1, 2 and 3 points respectively by assessment. With this refinement, the hierarchy of the developed phrase map was expected to be more reliable because the suggested concepts and instances were more consistent with the experts' understanding of the discipline. After re-evaluating the term weighting in the entire phrase map by applying the scoring data assessed by the experts, the number of terms was reduced to 2,406, whereas the number of relationships decreased to 3,790. Furthermore, 407 terms were suggested as concepts and 2,146 as instances by the third weighting rule.

The final step in developing the ontology was to convert the concept-instance hierarchy into an OWL format. The OWL specification is the most popular ontology representation format endorsed by the World Wide Web Consortium (W3C) (Smith, et al. 2004). The primary sublanguage of OWL, OWL Lite, is properly used to support a classification hierarchy and simple constraints, and was thus selected for this research.

Discussions

This section presented a procedure for developing base domain ontologies by extracting knowledge from domain handbooks. The main aim is to use the base domain ontology to support the domain knowledge–based IR system, which will be introduced in the next section. However, the demonstration also provides a good reference to reduce the intensity of expert participation. In the procedure, terms and relationships are rapidly collected from the handbook's TOC, definitions, and index. A set of weighting rules was used to evaluate and revise the extracted phrase map. Finally, the developed base ontology is represented in the OWL Lite format to store the domain glossary and concept hierarchy information.

Although the base domain ontology developed could still be incomplete, the procedure provides a practical approach to quickly prepare a base domain ontology that can later be incrementally enriched. Especially, there are many existing engineering handbooks composed by similar important elements, such as glossaries, definitions, and table of contents that are organized by important concepts within the knowledge domain. In addition, the procedure is independent from the domain. As a result, the procedure can be applied to other engineering domains to develop their own ontology. With the help of computer algorithms developed in this research, the process of glossary development, the generation of hierarchies, the identification of term relationships, and the differentiation of concepts and instances can be automated. As a result, the participation of domain experts is minimized and their workload is limited to only ensuring the quality of the ontology developed. In addition, the presented approach is generally applicable to generate a base domain ontology for all engineering domains whenever similar handbooks are available.

ONTOPASSAGE: A NOVEL PASSAGE PARTITIONING APPROACH

This section will introduce how we apply the earthquake engineering base domain ontology to the NCREE reference collection to develop the domain knowledge–based IR system. Two important problems of technical documents—namely, their complicated structure and multiple concepts—were featured in the Introduction. Passage retrieval is one possible approach to address these problems. It partitions a large and multi-topic document into several "passages". The term "passage" does not necessarily refer to a paragraph in the document, but to any possible subset of text from the entire document. Passages can be automatically extracted from a document using existing algorithms (e.g., statistical passages and contextual passages which will be introduced in the next section) on the search engine end. After these passages are created, each passage can be treated as an independent document (Callan 1994) and becomes the smallest unit for IR indexing and searching.

We implemented passage retrieval as the core component of the domain knowledge–based IR system. We also designed a novel passage partitioning approach, called OntoPassage (Lin, et al. 2012), to generate passages, each with around 300 terms, according to domain knowledge.

Passage Retrieval

Partitioning a large document into passages not only leads to better IR effectiveness but also provides a better search result representation that is easier to read. Both help satisfy the information needs of search engine users more quickly. There are various approaches to partitioning a document. The lengths of the partitioned passages can be either fixed or variable. The passages can be categorized into Statistical Passages and Contextual Passages, depending on whether the partitioning considers the document's context.

Statistical Passages

According to some statistical characteristics, several thresholds, such as "text length longer than 50 words and shorter than 600 words", will be set as the criteria. The system partitions passages that match the thresholds. Most Statistical Passages are of fixed length, irrespective of the structure of the original document. Therefore, a passage may span more than two chapters or sections. Such a feature has the disadvantage that the IR effectiveness cannot be controlled. For example, if an important sentence is interrupted at the end of a passage, such a passage cannot be appropriately ranked because some information has been lost from the interrupted sentence. However, a fixed-length passage also has the advantage that the number of passages can be anticipated, and the algorithm is simpler for an engineer to implement.

Contextual Passages

The specifications for partitioning Contextual Passages are related to the context and structure of the document. Concepts or key terms inside a document or from other semantic resources will also be used to assist partitioning if available. Most Contextual Passages are of variable length. They can be generated according to the structure of the original document (Wilkinson 1994), such as the paragraphs, sections, or pages (Salton et al. 1993). Moreover, Contextual Passages can also be generated according to many pre-defined algorithms. For example, it can be generated by an instant query. The MultiText research group (Clarke et al. 2006) found that search engine users often use short terms of 2.9 words on average for their queries. Such a query term can be used to generate instant passages to match their information needs.

The Development of Ontopassage

As discussed in the previous section, Statistical Passages are easy to generate, and their data are smaller. However, such passages cannot focus on concepts, despite achieving better IR effectiveness than the original document. Statistical Passages cannot help readers understand how concepts are distributed in a document. Moreover, besides having their information needs met, users also hope to obtain related concepts of the information they are searching for from the documents retrieved (Barzilay and Lee 2004). For this reason, we chose contextual passages to develop the passage partitioning approach. Among contextual passages, the MultiText passage is the best prototype because it has concept-focused characteristics. Therefore, it was chosen as the foundation for the development of OntoPassage. The most important difference between the proposed OntoPassage approach and MultiText is that the latter is based on instant query terms (i.e. passages are generated instantly according to the query term input by users). But the search engine users have to wait for the passage partitioning process (after they input the query), whereas the former uses the base domain ontology (i.e. passages are generated in advance and so the search engine users do not have to wait for the passage partitioning process), which is discussed earlier, to replace the instant query terms.

To illustrate how a domain ontology can be used to assist passage partitioning, we present an example of a simplified domain ontology for the topic of "Structure Control" in Figure 1-6. Here, the concepts of Structure Control are represented in a hierarchical structure. There are seven concepts in this domain.

While the topic "Structure Control" is treated as a query term, the system will try to retrieve some key terms from the seven key concepts in the documents.

This study defines a passage generated by the approach described above as an "OntoPassage" because the boundaries of the OntoPassage are derived by a domain ontology.

The procedure for partitioning a document into OntoPassages can be divided into five stages, as shown in Figure 1-7. The first stage is to simplify the relationships of the base domain ontology extracted from the *Earthquake Engineering Handbook*. The second stage is to calculate the term frequency of each term in a document. Terms with a higher term frequency will be chosen as the main topics of the document. In the third stage, these main topics interact with the base domain ontology; several related concepts and relationships are adopted from the base domain ontology and form a "document ontology" for the document. In the fourth and the fifth stages, the document is partitioned into OntoPassages according to each main topic. That is, each topic will have its corresponding set of passages. In the fourth stage, positions will be marked where the related terms of a main topic appear. These positions will be candidates for passage boundaries.

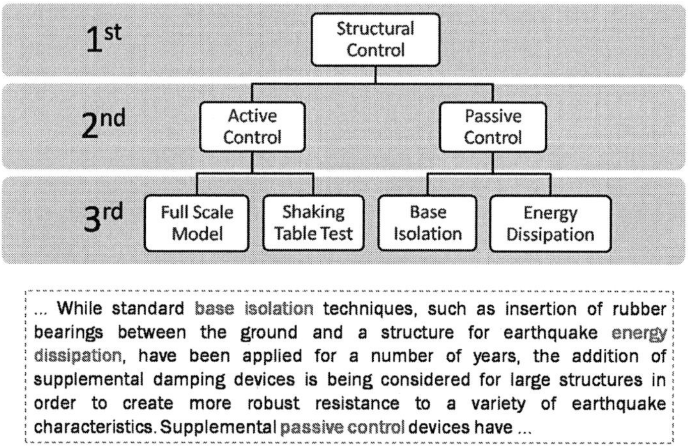

Figure 1-6. The concept of generating passages according to a domain ontology (Lin et al. 2012)

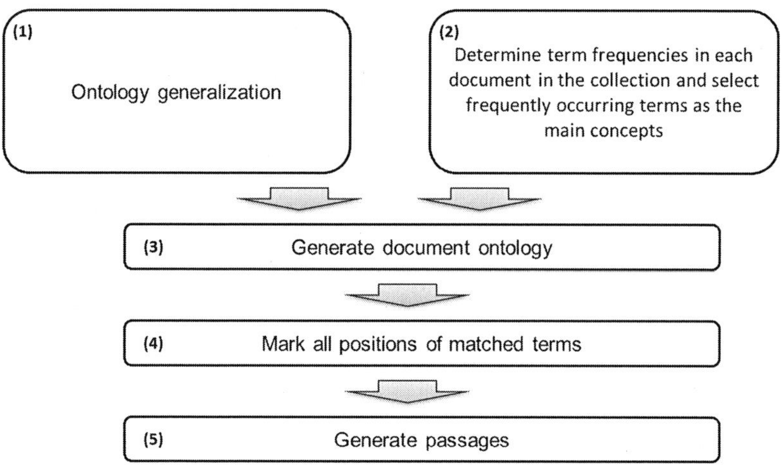

Figure 1-7. The procedure for generating OntoPassages (Lin et al. 2012)

Take Figure 1-6 as an example to explain Figure 1-7. Since we have the Earthquake Engineering base domain ontology (item No. 1 in Figure 1-7) and the NCREE reference collection (item No.2 in Figure 1-7), we can generate the document ontology (item No.3 in Figure 1-7) as the top side of Figure 1-6 shows. In Figure 1-6, we apply such document ontology to the document and find three key terms "base isolation", "energy dissipation", and "passive control". Then we mark the three positions of the three key terms (item No.4 in Figure 1-7) as the boundaries for the passage base units. Finally, we merge the two base units of passages to form an OntoPassage (item No.5 in Figure 1-7).

THE ONTOPASSAGE SEARCH ENGINE

System Development

The most important purpose of the proposed OntoPassage approach is to develop a domain knowledge–based search engine system. In a conventional IR scenario, the users of search engines have information needs that they must transform into appropriate query terms. The search engine system will then compare similarities between the query terms and the documents, and then retrieve the results for the users. Besides the rank of the documents, a search engine will also provide users with document titles and passages where the query terms appear. This information can help users judge whether the retrieved document matches his or her information needs.

However, if search engine users are not domain experts, they may sometimes have difficulties expressing their information needs in appropriate query terms. In this situation, displaying the concept map of a retrieved document may be quite useful. After the OntoPassage approach is introduced into a search engine, search engine users can select appropriate query terms on the concept map according to their information needs. The search engine then compares the similarities between the query terms and all of the OntoPassages. Such a concept-oriented scenario has two main advantages:

1. Search engine users can judge the relevance of a large document to their information needs by checking the OntoPassages retrieved, rather than having to review the entire document to find scattered passages.
2. When OntoPassages are generated, the domain ontology provides many additional related terms. If a search engine user does not know how to translate his or her information needs into appropriate query terms, such a procedure can expand the range of relevant query terms.

System Demonstration

Like most general-purpose search engine systems, the implemented system has three main user interfaces. The first is the main page of the search engine, which lets search engine users input their query terms. The second is the result viewer, which lists the retrieved documents together with their ranks and paragraphs of automatic summarization, and the third is the document viewer, which displays the content of a specific retrieved document if the search engine users click the link to the document.

In the concept-oriented retrieval scenario, the search engine users' point of view will shift to "concepts" instead of "terms", and the relationships between similar concepts will be displayed, as shown in Figure 1-8. The left side of the result viewer shows the documents retrieved. Under each document, the three passages that are most closely related to the concepts of the query terms are provided to users.

The right side of the result viewer shows the document ontology for each document. Users can therefore extend their search task using the linkages between different concepts.

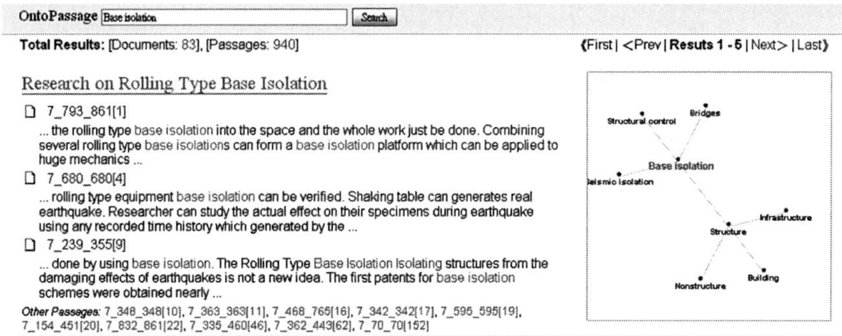

Figure 1-8. Result viewer of the domain knowledge–based search engine (Lin et al. 2012)

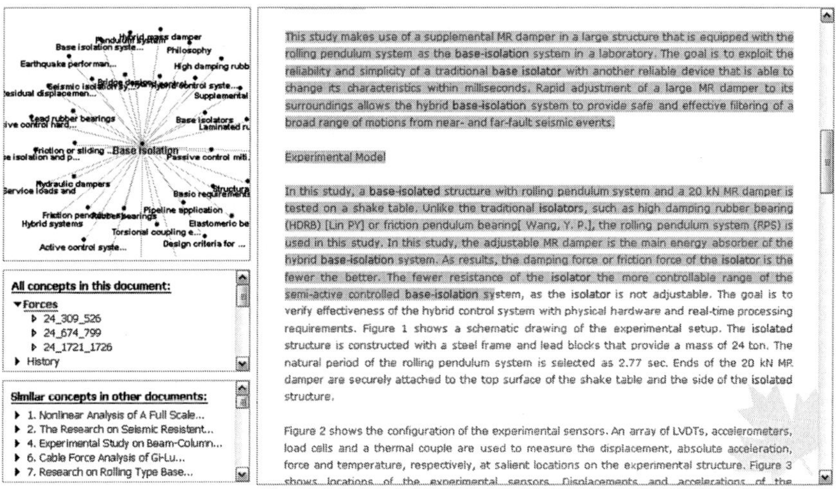

Figure 1-9. Document viewer of the domain knowledge–based search engine (Lin et al. 2012)

If search engine users want to browse a particular passage, the system will display the document that this passage belongs to and highlight the specific passage, as shown in Figure 1-9.

The left side of the document viewer is divided into three sections. The first displays the simplified document ontology to show the concepts that this document contains. The second also displays the concepts that appear in the document, and provides information on how many passages in the document contain that concept. Search engine users can switch the concepts that they care about (*i.e.*, the focused concepts) in this section. The last section displays how many other documents contain the focused concept. Therefore, search engine users can focus on one concept and retrieve it from the entire collection.

Compared with general purpose search engines, such a user interface can grab more terms and concepts from domain ontology. Like most general purpose search engines, the users must start by entering a query term by themselves. But the proposed domain-knowledge based search engine can link the query term to other relevant concepts, and also focus on specific passages which contain the important concepts within a document.

System Evaluation

As mentioned previously, the NCREE reference collection provides only a relevance assessment between the documents and the eight information needs. While each document is further partitioned into several passages, it is difficult for humans to judge the relevance of the passages to the information needs because of the very large number of passages. In other word, the information requests in the reference collection can be still applied. However, the relevance assessments no longer work since the documents are further partitioned into more passages (i.e., all the documents in the collection have changed. All the relevance assessment should be therefore re-performed). In order to evaluate the OntoPassage search engine system more precisely, we conducted a small-scale experiment to re-build part of the relevance assessments for the passages, and also to verify the system's effectiveness.

At first, the concepts of the eight information requests in the NCREE reference collection were treated as query terms and were sent to the system. The experiment was performed on both document retrieval and passage retrieval. The former returned a set of documents, and the latter returned a set of passages. The documents/passages were ranked by OKAPI BM25 model (Robertson, et al. 1996). The retrieved documents/passages were then evaluated by domain experts.

In both parts, domain experts were asked to answer similar questions about whether they could understand the essence of the documents or passages from the automatic abstract or the passage provided. They were then asked to perform two evaluations:

1. Evaluate whether the document/passage is related to a corresponding topic.
2. Correct the rank of each document/passage if they think the rank is inappropriate. They did not need to give a very precise rank as there were only three options: top 5, top 10, and lower than top 10.

After the evaluation, several IR indices could be further calculated. Table 1-4 shows the results of the assessments of the eight topics by the domain experts. In Table 1-4, "P@10" represents the precision of the top 10 documents/passages and "Rank Precision" represents the fitness of the ranking of the top 10 documents/passages.

These results demonstrate that the OntoPassage approach produces better search and ranking results than document retrieval, especially for Topics 2 and 10. In other topics, although the differences between precisions are not obvious, all topics had a better rank precision in the passage group. Therefore, users can obtain a more appropriate rank when they adopt the OntoPassage approach.

Table 1-4. Evaluations of the effectiveness of document retrieval and passage retrieval (Lin, et al. 2012)

		Document		*Passage*	
Topic No.	*Topic Name*	*P@10*	*Rank Precision*	*P@10*	*Rank Precision*
2	Seismic evaluation and retrofit	0.45	0.35	0.95	0.85
3	Seismic experiment	0.95	0.75	0.9	0.9
5	Structural control	0.8	0.6	0.7	0.7
6	Seismic hazard simulation	1	0.35	1	0.35
7	Computational mechanics	0.55	0.2	0.85	0.3
9	System monitoring	0.8	0.55	0.8	0.4
10	Ground motion	0.5	0.4	0.9	0.8
11	Geology and geotechnical engineering	0.95	0.65	1	0.75
Average		**0.75**	**0.48**	**0.89**	**0.63**

Case Example

For further understanding the use of the domain knowledge-based search engine, the authors illustrate a simulated scenario in this section. Earlier the NCREE reference collection was introduced. It is an expanding document collection that contains earthquake engineering technical documents. These technical documents are essentially the annual reports written by the earthquake engineering domain experts employed by NCREE. Such a technical report collection can serve as an internal knowledge base. When the employees in NCREE encounter any technical problem on their work, they can look up the document collection to see if there is any possible solution. For example, when an employee is looking for some technical reports related to the topic "structural control" as Figure 1-6 shows, he or she can first browses the document ontology (as the top side of Figure 1-6 and the right side of Figure 1-7 shows). While browsing, the employee may narrow down the information need from the topic "structural control" to "base isolation" because the employee is specifically interested in the technical documents related to "base isolation" as Figure 1-8 shows. In this situation, he or she can first find several passages related to "base isolation" within specific documents. After browsing the snippet of the passages, he or she can further browse the entire document as shown in Figure 1-9.

CONCLUSION

This chapter examined issues associated with the effectiveness of information retrieval (IR) in technical documents. We applied passage retrieval as a solution to develop a domain knowledge–based search engine system to implement a novel passage partitioning approach. We built the essential elements of the search engine (*i.e.*, the base domain ontology) step-by-step, and then demonstrated how they can be integrated in the system. Several typical passage-partitioning approaches were also reviewed, and the OntoPassage approach was proposed based on the MultiText approach. The OntoPassage approach uses a base domain ontology to anticipate the query terms that users may send to search engines, so that passages can be prepared in advance. It overcomes the problem of the MultiText approach's unsuitability to a large collection (*i.e.*, passages are generated instantly, so users need to wait for a system response).

The main purpose of OntoPassage is to provide a domain knowledge–based IR system. We developed a prototype search engine based on the OntoPassage approach and demonstrated it. Search engine users can use this to compare different IR scenarios of traditional document retrieval and ontology-based passage retrieval. Using the OntoPassage approach, search engine users can acquire extra knowledge and concepts when they perform IR tasks. In addition to the improved effectiveness of IR, the OntoPassage approach also makes good use of domain knowledge to generate meaningful and concept-focused passages. It can be concluded that, with the proposed OntoPassage approach, the IR effectiveness for domain-specific technical reports, as well as the presentation of IR results, can indeed be improved. A lesson learned from this research is that such a domain knowledge-based IR approach needs to be supported by a domain ontology. As a result, the quality of the ontology may also affect the IR effectiveness. Such a limitation should be noted because it might be more challenging for some domains to derive their own ontology. In the future, the authors will aim to enhance the quality of domain ontologies and improve the user-search interaction interface.

ACKNOWLEDGEMENTS

We would like to thank the Advanced Engineering Informatics journal for authorizing the reuse of the figures, tables, and parts of the contents from the papers (Lin, et al. 2008; Hsieh, et al. 2011; Lin, et al. 2012) we have published in the journal.

References

Barzilay, R. and Lee, L. (2004). "Catching the Drift: Probabilistic Content Models, with Applications to Generation and Summarization." in HLT-NAACL 2004: Proceedings of the Main Conference, Boston, MA, USA, May. 2–7, 2004, pp. 113–120.

Callan, J. P. (1994). "Passage-level evidence in document retrieval." in *Proceedings of the 17th annual international ACM SIGIR conference on Research and development in information retrieval.* Dublin, Ireland, Jul. 03–06, 1994, pp. 302–310.

Castells, P., Fernandez, M., and Vallet, D. (2007). "An Adaptation of the Vector-Space Model for Ontology-Based Information Retrieval." *IEEE Trans. on Knowl. and Data Eng,* 19(2): 261–272.

Chen, W. F. and Scawthorn, C. (2003). *Earthquake Engineering Handbook.* Boca Raton, Florida, CRC Press.

Clarke, C. A., Cormack, G., Lynam, T., and Terra, E. (2006). "Question Answering By Passage Selection." *Advances in Open Domain Question Answering,* 32: 259–283.

Diraby, E., Lima, C., and Feis, B. (2005). "Domain Taxonomy for Construction Concepts: Toward a Formal Ontology for Construction Knowledge." *Journal of Computing in Civil Engineering,* 19(4): 394–406.

El-Diraby, T. E. and Kashif, K. F. (2005). "Distributed ontology architecture for knowledge management in highway construction." *Journal of Construction Engineering and Management,* 131(5): 591–603.

Garcia, A. C. B., Kunz, J., Ekstrom, M., and Kiviniemi, A. (2004). "Building a project ontology with extreme collaboration and virtual design and construction." *Advanced Engineering Informatics,* 18(2): 71–83.

Gruber, T. R. (1993). "A Translation Approach to Portable Ontology Specifications." *Knowledge Acquisition,* 5(2): 199–220.

Hsieh, S. H., Lin, H. T., Chi, N. W., Chou, K. W., and Lin, K. Y. (2011). "Enabling the development of base domain ontology through extraction of knowledge from engineering domain handbooks." *Advanced Engineering Informatics,* 25(2): 288–296.

Issa, R. R. A. and Mutis, I. (2006). "Ontology based framework using a semantic web for addressing semantic reconciliation in construction." in *Proceedings of the 13th international conference on Intelligent Computing in Engineering and Architecture.* Ascona, Switzerland, Jun. 25–30, 2006, pp. 348–367.

Lima, C., El-Diraby, T. E., and Stephens, J. (2005). "Ontology-based optimization of knowledge management in e-construction." *Journal of IT in Construction,* 10: 305–327.

Lin, H. T., Chi, N. W., and Hsieh, S. H. (2012). "A concept-based information retrieval approach for engineering domain-specific technical documents." *Advanced Engineering Informatics,* 26(2): 349–360.

Lin, K. Y., Hsieh, S. H., Tserng, H. P., Chou, K. W., Lin, H. T., Huang, C. P., and Tzeng, K. F. (2008). "Enabling the creation of domain-specific reference collections to support text-based information retrieval experiments in the architecture, engineering and construction industries." *Advanced Engineering Informatics,* 22(3): 350–361.

Mizzaro, S. (1998). "How many relevances in information retrieval?" *Interacting with Computers,* 10(3): 303–320.

Noy, N. F., Sintek, M., Decker, S., Crubézy, M., Fergerson, R. W., and Musen, M. A. (2001). "Creating Semantic Web Contents with Protégé-2000." *IEEE Intelligent Systems,* 16(2): 60–71.

Peachavanish, R., Karimi, H. A., Akinci, B., and Boukamp F. (2006). "An ontological engineering approach for integrating CAD and GIS in support of infrastructure management." *Advanced Engineering Informatics,* 20(1): 71–88.

Robertson, S. E., Walker, S., Jones, S., Hancock-Beaulieu, M. M., and Gatford, M. (1996). "Okapi at TREC-3." *Overview of the Third Text REtrieval Conference (TREC-3),* Gaithersburg, MD, Nov. 2–4, 1994, pp. 109–126.

Salton, G., Allan, J., and Buckley, C. (1993). "Approaches to passage retrieval in full text information systems." in *Proceedings of the 16th annual international ACM SIGIR conference on Research and development in information retrieval.* Pittsburgh, Pennsylvania, USA, Jun. 27–Jul. 1, 1993, pp. 49–58.

Sheskin, D. J. (2003). *Handbook of Parametric and Nonparametric Statistical Procedures: Third Edition*, Taylor & Francis.

Smith, M. K., Welty, C., and McGuinness, D. L. (2004). "OWL Web Ontology Language Guide." from http://www.w3.org/TR/owl-guide/ (Mar. 10, 2014).

Uschold, M. and Gruninger, M. (1996). "Ontologies: principles, methods, and applications." *Knowledge Engineering Review*, 11(2): 93–155.

Voorhees, E. M. (1998). "Variations in relevance judgments and the measurement of retrieval effectiveness." in *Proceedings of the 21st annual international ACM SIGIR conference on Research and development in information retrieval.* Melbourne, Australia, Aug. 24–28, 1998, pp. 315–323.

Wilkinson, R. (1994). "Effective retrieval of structured documents." in *Proceedings of the 17th annual international ACM SIGIR conference on Research and development in information retrieval.* Dublin, Ireland, Jul. 03–06, 1994, pp. 311–317.

CHAPTER 2

Wiki-Con: A Pragmatic Approach for Semantic Interoperability for Construction Projects

Ivan Mutis*
Raja R.A. Issa*

Abstract: *This research addresses semantic interoperability by conceptually exploring the activities of communication when individuals aim dissimilar commitments to concept meanings in construction projects (i.e., addressing multiple commitments to the meanings of the same object or, alternatively, multiple ways to define such object). This research proposes a conceptual approach based on relevance theory to conceptually define multiple independent knowledge contexts in projects, which are user, community of practices, and modeler contexts. The conceptual approach gives clarity on the way the commitments to the meaning of representations take place in construction projects. The approach is pragmatic since it is concerned with the importance of the project participant dynamic as fundamental conceptual elements. The approach defines ontological categorizations to analyze the project social fabric, which dynamic establishes its underlying mechanisms. The claim is that when the dynamic is captured by the use of mediation technology, it is possible to register the participants' positive cognitive effect and cognitive processing effort. Such effects and efforts are in terms of perception, memory, and inference, required to choose an input from the mass of competing stimuli. Therefore, the underlying mechanism mirrors the dynamic of social fabric for the participants, which translates into what is relevant to the participant within the context. To demonstrate this approach using mediation information technologies, this project built as semantic resource system, Wiki-Con. Semantic resources are all sort of artifacts with mediation meaning capability through sign systems. Wiki-con's functionality is akin to a knowledge broker system. At the core of the Wiki-Con system is domain ontology that focuses*

*Center for Advanced Construction Information Modeling; M.E. Rinker School of Construction Management, College of Design Construction and Planning, University of Florida; email: imutis@ufl.edu, raymond-issa@ufl.edu

on relating concepts from independent knowledge context rather than on building formal concept-correctness using formal linguistic-rigor. Wiki-Con serves as controlled vocabulary engine, which in turns provides class (concept) definitions for the construction project.

INTRODUCTION

Construction projects incorporate multi-disciplinary participants from different specialized fields including architecture, electrical, mechanical, and structural. These participants- or individuals- constitute a complex, social, heterogeneous, human network composed of a fragmented group of individuals whose teamwork results in the creation of collective entities in projects. These collective entities, however, are disjoint groups whose complexity and heterogeneity hinder their ability to share, integrate, and exchange information within their routines and practices. In fact, since project tasks in organizations are fragmented (Galbraith 2002), the large number of specialties makes it difficult to integrate project activities. For instance, when individuals participate and play roles within construction firms, their project organization specializations and technology sophistication levels result in highly fragmented actions and interactions, which hinder their ability to share information.

Fragmentation in the construction industry has dominated researchers efforts in the last decades, which have mainly focused on technology transfer among construction project participants (Fenves 1996). Recently the community has made important strides to address this problem through various initiatives that ranges from integration of existing users' computer environments (Fischer, et al. 1998, Luiten, et al. 1998, Viljamaa and Peltomaa 2014) to creation of semantic grids (Turk, et al. 2005) to standardization on Industry Foundations Classes (IFC) (BuildingSmart 2014).

This research takes a different perspective to tackle the effects of fragmentation on information sharing. It addresses the natural complexity conditions encountered in construction projects. Natural complexity refers to the uniqueness and dynamic of construction projects, which makes the effective coordination of workforce activities and information sharing difficult. For instance, varying local conditions in a physical context (uniqueness) is concomitant with high degrees of freedom in worker choice of actions (dynamic) in a social context. These features are translated into a complexity of construction project contexts (physical and social). Therefore, this research investigates methods and strategies to reduce the fragmentation effects within the complexity of project contexts for more effective information sharing activities.

This research proposes a pragmatic approach to facilitate information sharing activities. The pragmatic approach elaborates on the notion of context-independent knowledge by addressing how project participants and members of the Architecture, Engineering, and Construction (AEC) domains make commitments for their meanings. The approach enables participants draw meanings from independent knowledge contexts through mediation information technologies. To understand the complexity and the framework of the approach consider the following explanation of the construction project natural complexity in information management, which leads to the semantic interoperability examination in the AEC domain.

Semantic Interoperability

The ability of employing electronic mediation instruments for exchanging, sharing, and integrating information in a meaningful way where all parts functionally work together is known as semantic interoperability. Two important aspects are critical in semantic interoperability (Mutis 2012): (1) the used information, symbolized by representations such as visual and textual based representations, and (2) the meaning commitments from different individuals.

Semantic interoperability research is aimed at sharing common vocabulary and models among project actors. These efforts embrace the development of common, shared models and construction industry standards. Examples are IFC (BuildingSmart 2014), BS6100 (BSI 2010), OmniClass (CSI 2012). The objective is that multiple construction participants ultimately recognize and easily manage the shared models and set a universal language. The implementation and use of the models and the common vocabulary provides the possibility of reusing the information and of facilitating the integration of information within applications and data models. However, having a commonly shared model (i.e., a universal language) for information sharing and integration is an assumption, which is based on a consensus and an understanding of commitments on meanings.

Conversely, the capability of common shared models to semantically map the information content is hampered by the lack of mechanisms to incorporate connections to the interpreters and information purpose from the source. The industry standards and common vocabulary overlook the relationship between the interpreters and the shared representation for a consistent interpretation. They lack mechanisms to connect the knowledge engineers and interpreters, which leaves gaps when interpreting the shared information. The creation of standards and common models ignores the communities of users and sidesteps issues that may explain the information purpose. In consequence the association from experts' knowledge to the corresponding standards discounts the community of users' purpose for using the information. Therefore, research approaches that lead to designing mechanisms for semantic reconciliation from individuals (experts), modelers, and members of the communities of practice are required. Research on the foundations of meaning commitment dynamics is of particular interest.

PRAGMATIC APPROACH FOR SEMANTIC RECONCILIATION

The approach addresses the dissimilar commitments to concept meanings, since one or more participants make different ontological commitments to the same concepts. For instance, project participants assign meanings to shared objects differently within a project activity (social dynamic activity), which in turn implies having different views of the object. In consequence, there are multiple commitments to the meanings of the same object or, alternatively, multiple ways to define such object. In linguistics, two main theories of communications address this dynamic.

The first is speech act theory, which focuses on decoding a message by deciphering the assigned meanings to utterances. This theory provides a characterization of linguistic expressions. The speech act defines rule-governed forms of behavior when individuals play a role in communicating information through language, specifically through the act of speaking. There are set of sufficient and necessary conditions for the speech act performance where kinds of behavior, such as intentional behavior, can be characterized (Searle 1985). The speech act characterizes what the speaker (sender) communicates to the hearer (receiver) by relying on the mutually shared background of the information or contexts and the intention of the utterance (Searle 1969).

The second is relevance theory based on inferential pragmatics. Advanced by Wilson and Sperber (2004), the focus is on inferential models to answer what sort of utterances and observable phenomena may be relevant. These experiences provide an input to cognitive processes and inform relevance to an individual at some point in time. Important here are the available assumptions related to context, which yield to a *positive cognitive effect* to infer conclusions worth having (Sperber and Wilson 1995). The *cognitive effect* is a *contextual implication* driven by the input and the context together, but from neither input nor context alone (Wilson and Sperber 2004). Relevance theory claims a level cognitive processing effort in terms of perception, memory, and inference required to pick out an input from the mass of competing stimuli. The more effort required processing the input, the less rewarding and relevant is effort.

By drawing from these precepts for semantic reconciliation, the defining factors on relevance are the foci to define contexts (e.g., relevance on any construction project participants' utterances and observations). These factors on relevance are the positive contextual effects yielded by (1) the available assumptions to the individual, and by (2) the required processing efforts for inferences from the input (i.e., processing efforts to make inference from observable models, or text-based and graphical representations as input). Therefore to analyze these effects, independent knowledge contexts are defined for (1) users, or project participants–usually from the participating organizations in the project, (2) external members to the project, such as members the communities of practice, and (3) modelers, who give information through computer enabled systems to the users (see Figure 2-1).

For semantic interoperability - the ability of employing electronic mediation instruments in a meaningful way where all parts functionally work together-, the inferential pragmatic approach aims at analyzing and mapping the previous contexts. In contrast for semantic interoperability, research approaches analyze or decode (breakdown) models to execute mappings of such models and experts' views. The decoding approach analyzes models built upon modelers' and experts'

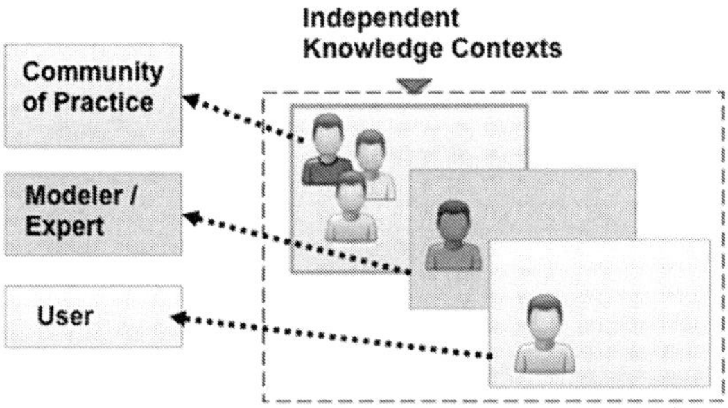

Figure 2-1. Contexts where factors of relevance take place

views, to reuse, map, integrate, or align their content, pieces of data, and semantics (e.g., mapping ontologies, conceptual models, and building common vocabulary, standards, etc.,). Decoding refers to the moments where an interpretation of sign systems occurs. In this approach and when using information technologies, research efforts are aimed at automatically or semi-automatically reusing, mapping, integrating, or aligning their content, pieces of data, and semantics. The relevance approach, however, is pragmatic. It addresses the dissimilar commitments to concept meanings when one or more participants make different ontological commitments to the same concepts. It is based on context-based meaning as opposed to analyzing modelers' or experts' views. The decoding model approach is based on analyzing syntax (text, html tags) and structure-based meanings (the semantics that the structures refer to – e.g., XML tags references) built upon the modelers' and experts' views.

Therefore, this research advances semantic interoperability by exploring methods for inferential pragmatics or context-based meaning capture in the AEC domain. It addresses the dissimilar commitments to concept meanings when employing electronic mediation technology to find ways where all parts work together. For this purpose and drawing from the inferential pragmatic approach, three main ontological categories are defined for social actors in the domain (i.e., any participating social individual in the project): users, community of practice members, and modelers. These ontology categories are fundamental layers since they are the defining frames for independent knowledge contexts in the domain.

The social aspect in this approach is fundamental which leads to the study of the participants' social dynamic (e.g., project participants' dynamic from different disciplines). In turn, the social dynamic is the enabling factor that articulates the independent contexts (i.e., one can define how one independent context is related to each other by establishing the social dynamics). The participants' social dynamic in the context also articulates the context components (i.e., one can define how the context components are related to the other by stating the participants' social dynamic. Thus, this research defines ontological categorizations to analyze the social fabric and in turn its dynamics. The following are three main ontological categorizations to interpret the participants' dynamic for this purpose: (1) entity, (2) activity, and (3) role. For the independent knowledge

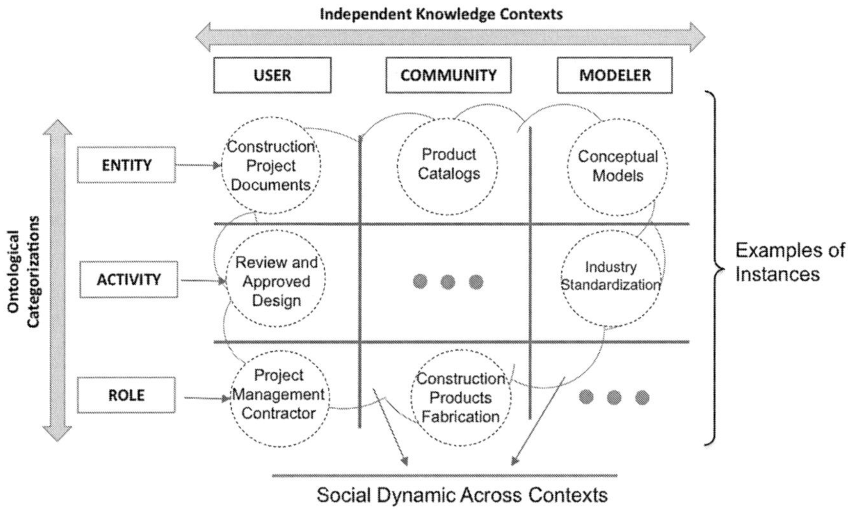

Figure 2-2. Intersecting Ontological categorizations and independent knowledge contexts by the participants' social dynamic

context analysis, for example, building ontological categorizations for participating social actors and purposes (i.e., building relation such as user-entity, user-activity, and user-role) enables further specializations levels to naturally represent the participants' social dynamic. Examples of the resulting instances of the ontological categorizations are shown in Figure 2-2.

As such, the three the independent knowledge context (user context, community of practice context, and modeler context) incorporate the participants' social dynamics including the ontological categorizations.

Independent Knowledge Contexts

Having the previous three independent knowledge contexts implies an informative contextual meaning generation (i.e., context-based meaning capture) based on each one of the contextual assumptions. The participant's dynamic incorporates underlying mechanisms that, if captured, register the positive cognitive effect and cognitive processing effort. Such effects and efforts are in terms of perception, memory, and inference, required to choose an input from the mass of competing stimuli. The underlying mechanism mirrors the dynamic of social fabric for the participants within the independent knowledge context, which translates into what is relevant to the participant within the context.

Wiki-Con, as Technical Implementation of the Pragmatic Approach

Wiki-Con functions as the underlying mechanism. Wiki-Con connects the three independent knowledge contexts by incorporating relevance for each context into an ontology-based system. Wiki-Con functionality is akin to knowledge broker system. Wiki-Con plays an enabling and supporting process of sharing independent knowledge and bridging contexts for the project participants. By incorporating relevance, project participants (users) are better able to capture assumptions and purposes made by others by creating a conductive environment enabled for decision-making. The mechanism captures the effects and level of effort as pragmatic instances.

Critical, therefore, is that any information system built to address semantic interoperability should be capable of capturing views in a given representation so that it conveys the necessary and sufficient conditions to enable individuals to understand any given meanings. However the design and implementation of such system is a major challenge. It implies that the system should have the capability of using techniques where the user is able to associate mental concepts to the representations. These mental concept associations were famously represented by the Ogden and Richards (1930) semiotic triangle. Wiki-Con establishes some strands to contribute towards this challenge. It creates a mechanism to bring the representation instances from three independent knowledge contexts. For example, when dealing with the community of practice context, Wiki-Con performs an analysis based on natural language processing. When dealing with user's context, as the user interacts with the system, Wiki-Con captures the frequency the user search for (query) instances or classes in the project representations at the user level. This mechanism associates the individuals' modus operandi with the representation so it leads to recognition of the user's approach to the representation, which in turn gives semantic clues for the users-representation relation. Wiki-Con associates the resulting instances (semantic clues) as values to the concepts registered in the core ontology, which add meanings to the ontological concepts. Representations are treated as instances that add semantic values to the core ontological concept.

WIKI-CON: A SEMANTIC RESOURCE SYSTEM

Wiki-Con is semantic resource system with functionality akin to a knowledge broker system. At the core of the Wiki-Con system is domain ontology for the architecture and construction domain, which focuses on relating concepts from independent knowledge context rather than on building formal concept-correctness using formal linguistic-rigor. This research subscribes to the definition of ontology as a formal and explicit specification of a shared conceptualization of a domain of interest (Gruber 1993) using artefacts, whose representational units are intended to designate some combination of types, classes, and relations between them (Klein and Smith 2010).

Wiki-Con serves as controlled vocabulary engine, which in turns provides class (concept) definitions for the construction project. The system outputs deliver the effects and level of effort for the independent knowledge context (user, community of practice, and modeler). The outputs are pragmatic instances, which are the ontological commitments drawn from the independent knowledge contexts and represented through instances. Thus, the outputs are ontological instances (general concept and relations in the domain at the modeler context level), and community of practice and ad-hoc instances (instances and relations at the context user level).

Concepts (or classes) and relations are two basic parts of the Wiki-Con domain ontology. For clarity and under the semantic interoperability inquiry area, the following section discusses some concept positions used for the Wiki-Con system construction. This clarification is fundamental since Wiki-Con is built on pragmatic stands.

Concepts and Instances

Concepts refer to the individuals' commitment to the meaning of a representation as a result of a cognitive process. Since semantic interoperability in a pragmatic approach is the concern of this research in the AEC domain, the individuals' understanding of the electronic mediation for exchanging, sharing, and integrating information in a meaningful way is critical.

The following section includes the explanation of how instances and concepts are built in the Wiki-Con system to bring clarity to the ways the commitment to the meaning of the representation is executed in the approach.

Instances for the AEC domain refer to entities, individuals, schemas, scripts, structures, including processes that exist in space and time. Thus, the content of project representations such as drawings and specifications are instances of project information. It is important to note that the content is mediated through syntax or text based systems of signs in the system. These instances stand to each other in several multilevel relations (Smith 2004). Understanding the effect of this fundamental multilevel relation is critical (e.g., one term contains the other forming a collection of terms). This means that it is possible to define collection of instances and associate the defining terms to such collections at a given time (e.g., every *metal_access_door* for *openings_building_rooms*). It is also possible to associate a class to define a collection if a term comprehends all instances from the collection of its members (e.g., *metal_door* class, as the collections of all doors that are made of metal). Thus, there are two possible designations when collections of instances are refered to. The first designation is *type_of,* where the collection of instances refers to all possible instances that the defining term instantiate at a given time (e.g., *doors* instantiate the collections of *revolving_doors*, thus we have *revolving_doors* as *type_of* door). The second designation is the collection of instances (not by type) associated by some features at given time (e.g., *temporary_building_windows*). For the second designation, the term refers to a collection of instances that comprehend all and only the entities where the term applies. For example, *temporary_builiding_windows* refers to all instances of the building in a construction phase, where access between two environments is possible. However, the *temporary_builiding_windows* term does not refer to the corresponding *metal_windows* type of *metal_made_window*.

The term class is commonly used instead of the term concept to explicitly refer to the collections of instances where a corresponding term applies (Smith 2004). For example, class *wood_doors* (general term) holds for all instances of wood made doors (the collection where the physical correspondences were agreed upon at any given time). Therefore from here and forth and for clarity, this work considers concept as classes.

Thus, the Wiki-Con core ontology contains classes where instances and types include the given general term (e.g., blue-door represents the collection of doors as openings which have an extension of being blue). Wiki-Con recognizes the dialogical character (dynamic social character) of the representations (any instance of the representation) where individuals interact within the independent knowledge contexts. Wiki-Con enables the associations of instances from the participants' independent knowledge context to the representations so some context conditions at the user, participants, and modeler level are captured.

Relations

Classes are ontological and they are different than the ones defined in databases or in knowledge bases. Although there are parallels in the treatment of classes, there are fundamental differences. Thus, here ontological classes are defined as the individuals' commitment to the meaning of a representation, the meanings of which individuals in a community agree upon. Modellers and knowledge engineers define classes in databases and knowledge bases according to their individual or work group views. To explore other differences further, the relations (ontology-relations) among ontological classes are first explained.

Nodes represent ontological classes when the ontology is designed in information systems, and links represent ontology-relations between those ontological classes. These two elements classes and relations form an *ontology*. Examples of ontological relations are *is_a*, *part_of*, *located_in*, *derived_from*, and *has_participant*, among others.

Ontological relations narrow or delimit the meaning from one class to the other (e.g., glass is *part_of* window). Ontological relations hold some form of universality in the meaning (i.e., an assumption is drawn on the meaning when practitioners or group of individuals agreed upon it). Associative relations such as *part_of*, for example, should hold the same meaning between classes (e.g., *x part_of y*). For example, *every* knob is *part_of* a door concept.

However holding meanings implies universality in the definitions, which enrich the complexity of any validation of an instance or type in ontologies. This is the case of contingent inclusion. The level of complexity increases in architecture and construction disciplines since ad-hoc situations pervade, rather than disciplines such as biology where universality in the definitions are backed by scientific theories. To alleviate this complexity, this research incorporates the time variable. This variable is characterized as continuant (when an instance or class identity is recognizable over some extended interval of time) and occurrent (when an instance or class that lacks a stable identity during any interval of time) (Sowa 1999). Thus, when the meaning of Wiki-Con classes remains the same for a stable interval of time, those classes are continuant and are close to a more universality conditions. For example, the defined term instantiates the corresponding term at a given time (e.g, wood is *type_of* material, thus all instances of wood refer to a category of material).

It is assumed that classes for the AEC domain are easily mapped to some instance in the real-world (e.g., blue-door *is_a* opening). Mapping to the real-word refers to the term correspondence by *extension* to reality. However, this mapping to reality between instance and a class may be inconsistent (e.g., every instance of blue-door *is-an* opening does not hold), since such correspondences are not universal. Thus any relation to entities in the real world should not be used in place of the *type_of* or universal relations. Not all instances in the real world hold for a *type_of* relations (e.g., every blue-door is a *type_of* opening does not hold, because the instance of the class door *is-a* decorative door). In particular, this is critical since ad-hoc definition pervade in the AEC domain. Wiki-Con correspondences are *is_a* kind of relations to avoid further inconsistencies when classes are mapped into the core ontology. *Type_of* ontology relations are only used when experts (modelers) validate the effect of such instance in the relation in the core ontology.

CORE ONTOLOGY APPROACH

As the technical implementation of the pragmatic approach, Wiki-Con is the underlying mechanism to connect the three independent knowledge contexts by capturing the effect on relevance for each context. Wiki-Con uses a core-ontology for this purpose. The core ontology was semi-automatically built based on two phases: the first was on the existing domain data models' examination and selection as is the main modeler input, and the second was on the design methods to capture the user and community of practice input.

Phase I: The Modeller Input

Data models in the AEC discipline capture the modeler's independent knowledge context by incorporating the modeler view. The selected model should possess basic consensus on common vocabulary and commitments to the domain concept meanings, including a basic ontological structure. Next a review of ontology construction efforts in the AEC domain is presented since these research efforts also aim to build a basic ontological structure, as follows.

Conceptual Model Selection as Modeler's Input

The IFC model was selected as the main data model, as its latest versions comprehend a general and wide ranged of knowledge base for the AEC disciplines. Building the core ontology upon IFC results in the modelers' view transformation into a semantically rich formalization. The modeler independent knowledge context is in consequence formalized. This formalization is the main relevant input from the modeler's independent knowledge context. The formalization constitutes the Wiki-Con core ontology.

Wiki-Con ontology was structured in OWL. For instance, the initial formalization using IFC, defined in an EXPRESS schema with more than six hundred classes, was performed using direct mapping from Express Entities to OWL classes and Express attributes to OWL slots. The formalization was manually performed including manual class check for quality check on the mapping accuracy. These manual modifications were performed when required in the OWL classes.

At the schema level, an example of the EXPRESS schema and OWL schemas mappings is shown in Figure 2-3. The Express building blocks are Entity Types, Sub-Entities, Super-Types, Properties, and Restrictions. Direct mapping from EXPRESS entities to OWL classes, as well as EXPRESS Sub-Entities to OWL Sub-Class were performed. OWL cardinality was used to refer to the EXPRESS attribute's optional flag. Since OWL has also data property and object property, they were used to represent EXPRESS simple type attributes (boolean, integer, string), while the object property was used to represent user-defined attributes such as in OWL class. Enumeration Data Types, which define a set of names for one type in a domain, were declared to be extensible so that they could be extended in other schemas and were mapped to the *One_Of* enumeration classes in OWL. Select Data, which defines a choice or an alternative between different options, was mapped to *Union_Of* classes in OWL. Aggregation Types refer to list, array, set and bag in EXPRESS. However, there are no equivalent OWL structures to represent aggregation types. To overcome this limitation, a transformation was performed by defining an intermediate class to represent the aggregation type in an OWL class, so that the contained classes are a subclass set from a larger class. The purpose was to easily control properties and restrictions of each type using a larger class. In the same way, it was not possible to translate EXPRESS rules into OWL since OWL can only define property and its range, but not constraints that are dynamic properties. EXPRESS defines a couple of rules how a datatype can be further specialized. This is important for re-declared attributes of entities. For example, WHERE is a rule in EXPRESS, it must be evaluated into TRUE, otherwise a population of an EXPRESS schema, is invalid. Figure 2-3 shows an example of the mapping of IFC data to the OWL class.

EXPRESS schema	OWL schema
simple types : INTEGER	XSDint
simple types : REAL	XSDfloat
simple types : BOOLEAN	XSDboolean
simple types : STRING	XSDstring
simple types : LOGICAL	XSDint (-1, 0, 1)
ENTITY	owl:class
SUBTYPE	owl:subClassOf
simple attribute type	owl:DataProperty
general attribute type	owl:ObjectProperty
ENUMERATION OF	owl:oneOf
SELECT	owl:unionOf
LIST, ARRAY, SET, BAG	owl:class

(a)

IFC class IfcBuilding in the EXPRESS language

```
ENTITY IfcBuilding
  SUBTYPE OF (IfcSpatialStructureElement);
    ElevationOfRefHeight : OPTIONAL IfcLengthMeasure;
    ElevationOfTerrain : OPTIONAL IfcLengthMeasure;
    BuildingAddress : OPTIONAL IfcPostalAddress;
END_ENTITY;
```

Jena code:

```
OntClass Building = m.createClass(IfcNS+"IfcBuilding");
Resource SpatialStructureElement = m.createResource(IfcNS+"IfcSpatialStructureElement");
Building.addProperty(RDFS.subClassOf, SpatialStructureElement);
Resource restriction = m.createResource(OWL.Restriction);
restriction.addProperty(OWL.minCardinality, "0", XSDDatatype.XSDint);
ObjectProperty                ElevationOfRefHeight                =
```

(b)

Generated OWL class:

```
<rdf:RDF
xmlns:rdf="http://www.w3.org/1999/02/22-rdf-syntax-ns#"
xmlns:owl="http://www.w3.org/2002/07/owl#"
xmlns:xsd="http://www.w3.org/2001/XMLSchema#"
xmlns:rdfs="http://www.w3.org/2000/01/rdf-schema#">
<owl:Class rdf:ID="IfcBuilding">
<rdfs:subClassOf>
<owl:Restriction>
<owl:onProperty>
<owl:ObjectProperty rdf:about="BuildingAddress"/>
</owl:onProperty>
<owl:onProperty>
<owl:ObjectProperty rdf:about="ElevationOfTerrain"/>
</owl:onProperty>
```

(c)

Figure 2-3. OWL class generation. Independent knowledge contexts

In sum, the modelers' input is the IFC data model current version (2.4), formalized into OWL which ontology and set up in the Wiki-Con system as the core ontology. The core ontology was built by parsing the IFC data schema through flat text-based EXPRESS language, then generating the ontology classes one by one for each IFC class using the Ontology Web Language (OWL)(W3C 2013) with Jena (2013) framework, a Java framework for building Semantic Web and Linked Data applications.

Phase II: The User and Community of Practice Input

The users perform manual operations as input of knowledge source to incorporate independent context for the user and the community of practice. The resources mirror the dynamic of social fabric for the participants within the independent knowledge contexts, which translates into the participants' relevance within the user and community of practice knowledge contexts.This view registers the continuous participants' action as the view captures contexts for the project through semantic resources used in project practices.

Wiki-Con incorporates semantic resources from the independent context as inputs. The user-context inputs are selected resources for the project that capture the dissimilar commitments to concept meanings, while participants incorporate such resources in their activities. Semantic resources that belong to the user knowledge context, for instance, are construction document specifications, which are created exclusively for the project. Currently Wiki-Con can process construction project documents based on the existing document structure such as OmniClass (CSI 2012). The purpose is to associate the document content to a formalized semantic structure, using syntax matching algorithms.

The system processes and stores the semantic resources in the Wiki-Con database. The Wiki-Con repository (i.e., the repository is the available semantic resource document management system) keeps records and updated document versions (processed specification project documents). The system processes each resource by creating a highly structured machine-readable file.

Thus, users select project documents and store them in the Wiki-Con database for further processing. There is a structured machine-readable file for each processed semantic resource.

Wiki-Con also processes dictionaries and taxonomies as semantic resources, whose domain is the community of practice independent knowledge context. By incorporating the semantic resources, Wiki-Con registers the cognitive processing effort of the user and community member, and embraces consensus or the community goals.

Associating Semantic Resource to the Core-Ontology

The associations arise from the independent knowledge contexts through the semantic resources. They extend from resources to the corresponding concept in the ontology. These associations are ontology instances. Wiki-Con uses an intermediate layer as machine-readable resources in RDF to link the resources

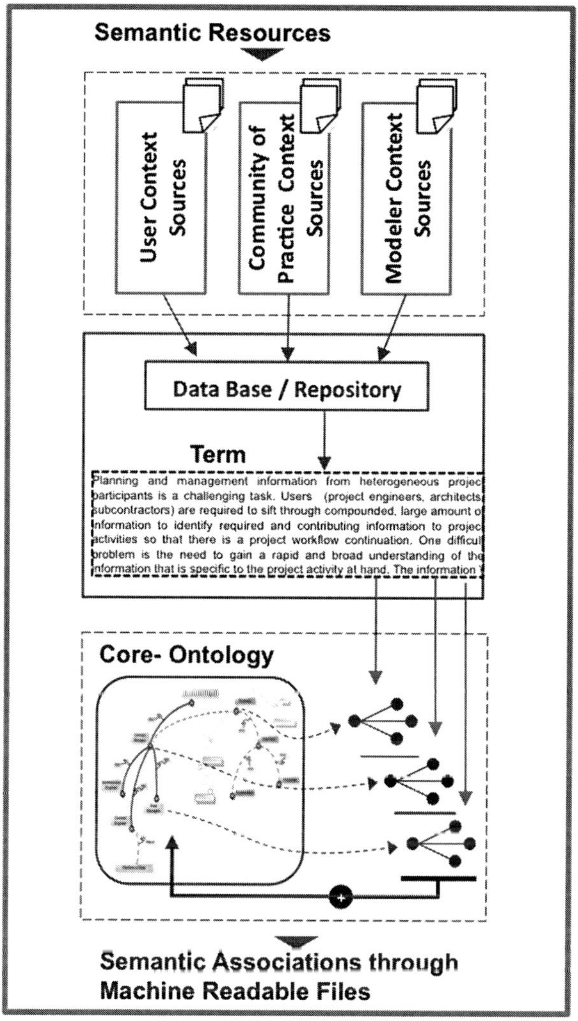

Figure 2-4. Knowledge extraction: associating semantic resources to the core ontology

and the ontology, which implies having semantically structured resources that are mainly the class terms (properties) and instances (multiple values).

As previously mentioned, Wiki-Con extracts knowledge from construction project documents based on the existing documents' metadata and data. The first key step is the extraction of data pertaining to the low level statistical properties of a construction document. This extraction requires parsing the document text from the database (See Figure 2-4). Thus, the extraction relies on statistics and not high-level analyses such as in natural language processing.

In particular, Wiki-Con creates a temporal machine-readable file using the structure, meta-data, and data for each available semantic resource. For example, the system extracts the related information from a Design Web Format (DWF) project document by reading the meta-data, the file structure, and the instantiations of such structures. CAD drawings are construction project documents in electronic format, typically expressed in DWF formats. The system further converts such resources in a formalized text-based document by extracting and organizing the metadata contained in the DWF files. Thus, each semantic resource has its associated structured machine-readable file and stored in the database. Currently, Wiki-Con capability is limited to DWF independent files and construction documents built on PDF formats based on a CSI structure, such as OmniClass (CSI 2012).

The knowledge extraction is based on the Core-Ontology classes (terms) frequency within the documents' data and metadata (See Figure 2-5). Thus, the assumption is that the repeated, specific term presence, which pairs in several different document sections, implies having a close relation between the terms. In turns, the assumption indicates that the terms found in documents are class instances from the independent knowledge contexts. Building on this assumption, Wiki-Con calculates the relationship strength between each pair of term based on a similarity metric. The relationship strength input is the frequency in which pairs of the terms occurs in the document. The pair, therefore, is the RDF term that corresponds to the core-ontology class name and the syntax matching terms in the semantic resource, e.g., terms within the construction specification documents' meta-data and data (text-based content). For this purpose, the Wiki-Con

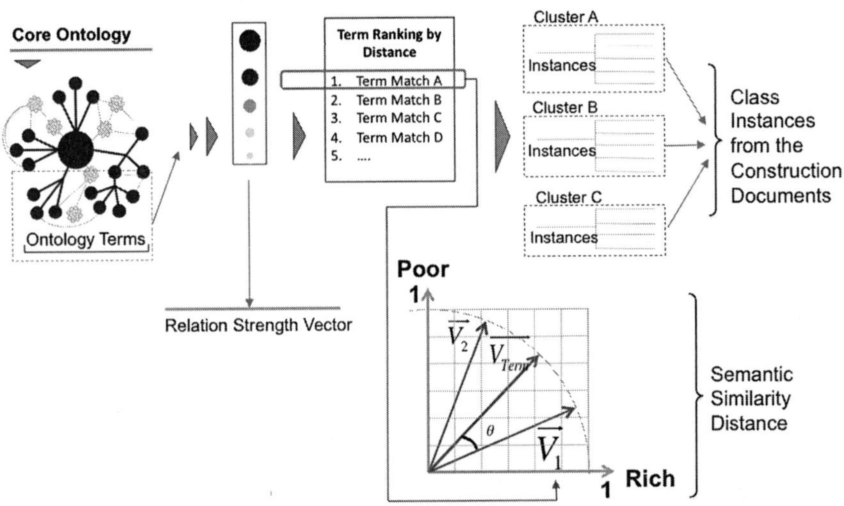

Figure 2-5. Instance association of semantic resources to ontology classes according to semantic similarity distance

uses *tf-idf* weights as statistical measure, and they are calculated using Equations 2-1 and 2-2:

$$w_{t,d} = \begin{cases} 1 + \log_{10} tf_{t,d}, & if \;\; tf_{t,d} > 0 \\ 0, & \text{otherwise} \end{cases} \qquad \text{Eq. 2-1}$$

$$idf_t = \log_{10}\left(\frac{N}{df_t}\right) \qquad \text{Eq. 2-2}$$

where, *tf* is the term frequency in document *d*, and *idf* is the document frequency for *t* or the number of documents that contain *t*.

The weight is a statistical measure to evaluate how important a term is to a document in a document collection. Therefore, the relevance increases proportionally to the number of times a term appears in the document, and it is offset by the term frequency in the collection of documents (Manning, et al. 2008). This weight is calculated using equation 2-3:

$$w_{t,d} = \log_{10}(1 + tf_{t,d,}) \times \log_{10}\left(\frac{N}{df_t}\right) \qquad \text{Eq. 2-3}$$

where, *w* is the weight and *N* the number of documents.

Based on the relationship term strength, Wiki-Con identifies the documents sections where the instance occurs. The instance is the ontological instance concept of the three intersecting independent knowledge context (user, modeler, and community of practice). Since, the number of instances may be significant, Wiki-Con creates a hash table for each ontological class name to register the relevant instance locations in the set of documents. The table is linked as RDF resource in the intermediate layer. The vector in the table holds results of the relevance test, as illustrated in Figure 2-5. The tables have indices and physical addresses of the extracted sections of documents for scalability purposes. Thus, users' searchers are conducted more efficiently since searching in the table reduces computational load and increase the search functionality, rather than having to manually parse the document contents of the entire database.

In sum, Wiki-Con uses syntax matching algorithms to map each semantic resource to the corresponding concept nodes (classes) in the core ontology, based on the strength of the relationship of the pair terms. The mappings are physical addresses to the data, metadata, or scheme, whose values are class instances in the core ontology. The purpose is to semantically enrich the core ontology with the associations through a controlled mapping process using the Jena (2013) framework reasoning rules. Although the mappings are automatically executed, the semantic resource is currently manually retrieved from the database.

Communicating Semantic Content and User Query Function

Considering the following example to illustrate the Wiki-Con utility for communicating semantic content.

Planning and management information from heterogeneous project participants is a challenging task. Project participants (project engineers, architects, subcontractors) sift through large amount of information to identify required and contributing information to project activities so that there is a project workflow continuation. One difficult problem is the need to gain a rapid and broad understanding of the specific information's relevance to the project activity at hand. The information should contribute to help plan and identify project activities based on the association of current and past information to the project (construction documents, catalogues, drawings, etc.). If technology mediates among heterogeneous sources, the challenge is how to allow users to relevantly examine the project information objectively and efficiently. Providing a rich data source to the user to comprehensively inform the project activities enables users to reduce the decision-making inefficiencies, such as invoking request for information and clarifications, or even avoid errors due construction documents misunderstandings.

By allowing users to query a class, Wiki-Con communicates the semantic content and the information relevance. When users query keywords, they invoke a semantic discovery process based on a syntax match in Wiki-Con. The required class is represented through syntactical terms (text-based) that have correspondences in the core ontology, so that the ontological meanings (definitions) enable decision-making for the planning or management of the project activity at hand. The query-function consists of a keyword search within the core-ontology. The search begins from the root class in the core ontology to build a similarity score vector, by matching a class or attribute. Thus, overall the search process has two phases: (1) finding the similarity (distance) between terms that characterize the relationship between the keyword and the ontology class; and (2) clustering information of the ontology class instance for the user's interpretation from a client browser through a Graphical User Interface (GUI).

The first phase involves finding correlations between two terms, the user's query keyword terms (a keyword may be composed of 1 to n words) and the occurrences of terms within the ontology classes. The correlation frequencies imply a close relationship between the terms. The purpose is to demonstrate through inferences the relevance of the existing class/conce instances, which are represented in project documents, so that the user will connect such inferences to enrich or define the user's understanding on the queried concept. Wiki-con output is the inferences and is based on similarity (distance measures). Wiki-Con uses the normalized Jaccard distance equation (Manning, et al. 2008), a version of the Cosine similarity distance equation, as follows.

$$\text{jaccard}(A,B) = \frac{|A \cap B|}{\sqrt{[A \cup B]}}, \qquad \text{Eq. 2-4}$$

where A and B are two data set vectors.

The second phase of the clustering process is to enable interpretation in the GUI, i.e., grouping the search results according to the relevance indicators ranked on distance values. The cluster divides large sets of terms (in this research case the terms refer to class instance names) into smaller groups, so that the smaller groups contain the closely related class instances. The clusters are organized according to a hierarchical structure to easily rank the classes of interest. A cluster example is the Wiki-Con output for interpretation, as shown in Figure 2-6 shown from the GUI. The Figure shows screens shots of the users' query 'texture' as a keyword term. It can be observed that the hierarchy syntax of the keyword 'texture' output cluster (ranked output showed in the GUI) is extremely similar. The class instances shown as text-based extractions, however, are semantic enrichments to the queried concept for the user's interpretation (for the user's decision making).

In Figure 2-6, the search not only shows the definition of the concepts through a weighting system, but also displays the class instances in existing project information, which in turns demonstrates the intersection of the independent knowledge contexts. The output is a highly controlled strategic visualization based on relevance ranking.

In sum, Wiki-Con assigns weights to classes and relations according to a relevance strategy to create a rank-based structure for the user. In principle, the ontology classes and its relations are relational structures. Thus, by using a weighing system, the relevance of a specific queried concept (class) should be

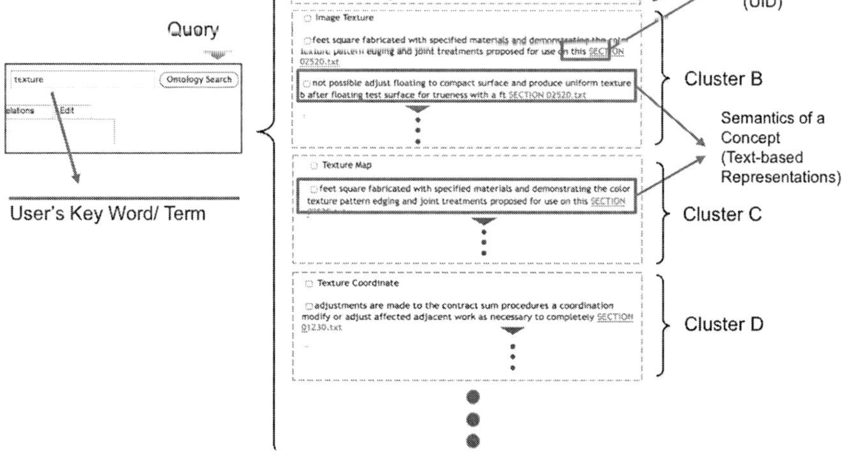

Figure 2-6. Wiki-Con user's query output: instances of existing project information

translated in the output. The weights are annotations to the relations and classes. The output is a set of the weightings that give semantic information about the queried class. The output in turn shows the relevance of such classes that arise from the independent knowledge context. Thus, weights are values and constitute logical constrains, which in turn are pragmatic-based constraints. Interesting to note is that the weighting system communicates the semantic context through a user query function, but the weighting system does not have any effects on the core ontology modeling or any influence in processing the semantic resources.

CONCLUSIONS

This research project focuses on the project participants' commitment to the meaning of the representations, as these representations are the mediation vehicle. Representations refer to all sorts of artefacts (artificial language, geometries drawings) which functionality is aimed at mediation on meanings that have been agreed upon by individuals such as project participants and members of a community of practice. Individuals at any independent knowledge levels use the representations as mediation mechanisms to express their commitment to the meanings, and these meanings are the ones that individuals agree upon. However when project participants commit to a meaning, they define such meanings on assumptions, which in turn are the individuals' view on which they make ontological commitments to the representation meanings. Furthermore, the individuals' view assumes that such ontological commitment is the same as the one developed by other individuals in the community of practice or in the project, which has direct implications on semantic interoperability.

While individuals have different views of what the representation means and implies, they also may agree on the meanings of some other representations but expressing and specifying the same representation meaning in multiple ways. In consequence the project participants' commitment to a meaning in activities of exchanging, sharing, and integrating information in a meaningful way, where all parts functionally work together, is a significant challenge.

Through the proposed pragmatic approach, this research advances semantic interoperability by conceptually exploring the activities of communication dynamics that address the dissimilar commitments to concept meanings (i.e., to address the multiple commitments to the meanings of the same object or, alternatively, multiple ways to define such object). The approach draws from relevance theory that is based on inferential pragmatics to conceptually define multiple independent knowledge contexts (user, community of practices, and modeler contexts). The conceptual approach gives clarity on the way the commitments to the meaning of representations take place. Having the proposed three independent knowledge contexts, where the commitments to the meanings take place, implies an informative contextual meaning generation based on each one of the contextual assumptions (i.e., context-based meaning capture).

The pragmatic approach underlies the importance of the project participant dynamic within the independent knowledge contexts as fundamental conceptual element, which is defined to articulate the context components. The approach defines ontological categorizations to analyze the project social fabric and in turn the fabric's dynamics. This dynamic establishes the underlying mechanisms that, if captured by using a mediation technology, it is possible to register the participants' positive cognitive effect and cognitive processing effort. Such effects and efforts are in terms of perception, memory, and inference, required to choose an input from the mass of competing stimuli. The underlying mechanism mirrors the dynamic of social fabric for the participants within the independent knowledge context, which translates into what is relevant to the participant within the context.

While the pragmatic approach elaborates on the notion of context-independent knowledge, by addressing how project participants and members of the AEC domains make commitments for their meanings, the approach facilitates the understanding of information sharing activities. It is anticipated that research efforts for semantic interoperability will be facilitated using this research approach framework. The approach is not only transformative in the AEC domains but also establishes a new course for research and development of mediation technologies where users draw meanings from independent knowledge contexts, which in turn serves as mechanism for the dynamic social fabric in the domains.

Therefore to further demonstrate this pragmatic approach using mediation information technologies, this project built as semantic resource system, Wiki-Con. Semantic resources are all sort of artifacts with mediation meaning capability through sign systems. Wiki-con's functionality is akin to a knowledge broker system. At the core of the Wiki-Con system is domain ontology that focuses on relating concepts from independent knowledge context rather than on building formal concept-correctness using formal linguistic-rigor. Wiki-Con serves as controlled vocabulary engine, which in turns provides class (concept) definitions for the construction project.

The Wiki Con system outputs deliver the effects and level of effort for the independent knowledge context (user, community of practice, and modeler). The outputs are pragmatic instances, which are the ontological commitments drawn from the independent knowledge contexts and represented through instances. Thus, the system outputs are ontological instances (general concept and relations in the domain at the modeler context level), and community of practice and ad-hoc instances (instances and relations at the context user level)

Wiki-Con also processes dictionaries and taxonomies as semantic resources, whose domain is the community of practice independent knowledge context. The purpose is to communicate additional semantic content for independent knowledge content. By incorporating the semantic resources, Wiki-Con registers the cognitive processing effort of the user and community member, and embraces consensus or the community goals. Although in its infancy, it is expect that Wiki-Con will evolve in a more robust semantic resource to express meanings from current representational vehicles in the AEC community.

References

BSI (2010). "Glossary of building and civil engineering terms." B. S. Institute, ed., 42.

BuildingSmart (2014). "Industry Foundation Classes (IFC)." *Developing open standards and specifications for BIM*, ⟨http://www.buildingsmart.org/⟩, (Accesed, April, 2014), Last Update 2010.

CSI (2012). "Construction Classification System, Omniclass." ⟨http://www.omniclass.org/index.asp⟩, (Accesed, January, 2014), Last Update.

Fenves, S. J. (1996). "Information Technologies in Construction: a Personal Journey." *Construction on the Information Highway* Z. Turk, ed., CIB Working Comission 78, 31–42.

Fischer, M. A., et al. (1998). "IT support of single project, multi-project and industry-wide integration." *Computers in Industry*, 35, 31–45.

Galbraith, J. R. (2002). *Designing organizations: an executive guide to strategy, structure, and process*, Jossey-Bass, San Francisco.

Gruber, T. R. (1993). "A translation approach to portable ontology specification." *Knowledge Acquisition*, 5(2), 199–220.

Jena, A. (2013). "Java framework for building Semantic Web and Linked Data applications." ⟨http://jena.apache.org/index.html⟩, (Accesed, April, 2014), Last Update March 2014.

Klein, G. and Smith, B. (2010). "Concept Systems and Ontologies: Recommendations for Basic Terminology." *Transactions of the Japanese Society for Artificial Intelligence*, 25(3), 433–441.

Luiten, G. T., et al. (1998). "Project-modelling in AEC to integrate design and construction." *Computers in Industry*, 35(1998), 13–29.

Manning, C. D., et al. (2008). *Introduction to information retrieval*, Cambridge University Press, New York.

Mutis, I. (2012). "Framework for Semantic Reconciliation of Construction Project Information." *Journal of Information Technology in Construction*, 17(1), 24.

Ogden, C. K. and Richards, I. A. (1930). *The meaning of meaning: A study of the influence of language upon thought and of the science of symbolism*, Harvest Books, Orlando, Florida.

Searle, J. R. (1969). *Speech acts: An essay in the philosophy of language*, Cambridge University Press, Cambridge, UK.

Searle, J. R. (1985). *Expression and meaning. Studies in the theory of speech acts*, Cambridge University Press, New York.

Smith, B. "Beyond Concepts: Ontology as Reality Representation." *Proceedings of FOIS*, IOS Press, 73–84.

Sowa, J. F. (1999). *Knowledge representation: Logical, philosophical, and computational foundations*, Brooks Cole Publishing Co, Pacific Grove, CA.

Sperber, D. and Wilson, D. (1995). *Relevance: communication and cognition*, Blackwell, Oxford, UK; Cambridge, Mass.

Turk, Z., et al. (2005). "Semantic Grid - Interoperability Solution for Construction VO?" *Proceedings of the International Conference on Information Technology: Coding and Computing (ITCC'05) - Volume II - Volume 02*, IEEE Computer Society.

Viljamaa, E. and Peltomaa, I. (2014). "Intensified construction process control using information integration." *Automation in Construction*, 39, 126–133.

W3C (2013). "Web Ontology Language (OWL)." ⟨http://www.w3.org/2004/OWL/⟩, (Accesed, April, (2010)), Last Update 2007.

Wilson, D. and Sperber, D. (2004). "Relevance Theory." *The handbook of pragmatics*, L. R. Horn, and G. L. Ward, eds., 27, 607–632, Oxford: Blackwell, Malden, MA.

CHAPTER 3

Ontology to Support Healthcare Facility Management

Jason D. Lucas*
Tanyel Bulbul†

Abstract: *Ontologies have become increasingly prevalent in healthcare for managing a wide range of information. They help to keep patient information and other records compliant with electronic data standards and, are also used, to a lesser extent, for the management of healthcare facilities' automated building control systems, sensors, and work order systems. Despite research that documents the importance of the condition of the physical environment for patient care and safety, a gap exists between the clinical and facility management systems within healthcare. They do not communicate or interact with each other to transfer valuable information needed for facility management's response to a critical event. In order to address this problem a healthcare facility information management system is proposed and designed with an ontology to help facility managers retrieve, track, and store relevant real-time clinical and facility management information. This chapter summarizes ontology efforts in both the healthcare and facility management industries. It also discusses the methodology used to design the healthcare facility management system, identified critical links between facility management and clinical information, and the developed ontology used to support facility management response to critical events.*

INTRODUCTION

With the advancement of Electronic Healthcare Records (EHR) in the healthcare industry ontologies have become a very important part of managing information and keeping patient information compliant with electronic data standards. Beyond EHRs,

*Department of Construction Science and Management, Clemson University, 136 Lee Hall, Clemson, SC 29634-0507; PH (864) 656-6959; FAX (865) 656-7542; email: jlucas2@clemson.edu

†Department of Building Construction, Virginia Tech, Bishop-Favrao Hall, 1345 Perry Street, Blacksburg, VA 24061; PH (540) 553-2140; FAX (540) 231-7339; email: tanyel@vt.edu

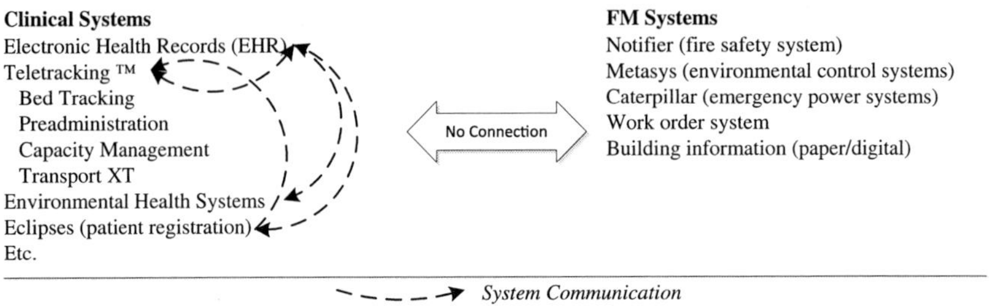

Figure 3-1. System disconnect

procedure schedules, medical equipment, bed allocations, medication systems, and nurses' stations all use electronic systems that exchange information within a complex web of defined communications. Incorporated into many healthcare facilities are building automation system sensors for controlling air conditioning temperature and humidity, security access, fire control systems, and other building control and automation systems including those used to manage maintenance work and work orders. These facility-based systems are used to support facility managers who are charged with the operation and maintenance of the environment used for clinical operations. The problem is that the two sides of technology within the healthcare facility, the former used for clinical service management and the latter used for facility management, do not interact and communicate proper information (Figure 3-1).

Within healthcare, research has shown that the condition of the physical environment is very important to patient safety, rate of recovery, and success of overall care (Dijkstra et al. 2006; Bakken et al. 2004). The upkeep to the facility and resolution to facility related problems has an impact on patient safety and clinical operations within the healthcare environment. The longer it takes facilities personnel to resolve an issue the greater chance of a patient safety event and greater risk to clinical operations exist (Lucas et al. 2013a). It is important that facility managers have the proper lifecycle information about the facility's past and current clinical happenings within the facility to ensure minimal impact on clinical operations while conducting repairs. In order to support effective and efficient response to facility related patient safety events an ontological framework has been proposed and designed to link the relevant real-time and historic clinical information with facility information (Lucas et al. 2011). This framework and how it uses real-time data and semantic reasoning to support facility management response is discussed in this chapter.

CURRENT LITERATURE

Ontology use within healthcare is quite popular and used for a variety of reasons from managing Electronic Health Records to context-aware computing in treating patients. With respect to Electronic Health Records (EHRs) ontologies have been used to keep them maintained in the proper format, updated to current standards, and transferable between systems using different standards (Rector et al. 2009; Iftikhar et al. 2010; Imam et al. 2007). Ontologies have also been proposed to help manage relevant information through a syntax-based semantic search engine (Rajan and Lakshmi 2012; Beyan and Baykal 2012).

The uses of ontology within healthcare have also been linked to patient safety and effectiveness of care. One effort is examining the use of ontology for creating a standard medical error identification system to improve future patient safety and preventing adverse events (Mokkarala et al. 2008). Ontology within an information infrastructure has also been analyzed for improving patient safety through automated surveillance for real-time error detection and prevention, increased communication, and standardization of practices (Bakken et al. 2004). A medical ontology has also been looked at to aid in emergency situations to help systems communicate pertinent information in the ever changing dynamics of a healthcare environment (Zeshan and Mohamad 2012).

Several efforts have also connected ontology to context-aware computing. These efforts include one used to help identify the best healthcare services for patients within a semantic web-based system (Fenza et al. 2012). This system was designed to take real-time information and through a combination of ontology and dynamic context (real-time) data help identify problems, manage situations with patients, and determine proper medical services. Similarly, ontology has also been used to help implement Clinical Guidelines, or defined procedures used to aid in decision making, by providing the sematic requirements needed for delivering proper care (Isern et al. 2012) and to help determine clinical pathways with semantic rules (Hu et al. 2012). This latter system took into account context data such as the illness, needs of the patients and clinical professionals, type of healthcare settings, medical guidelines, and resource limitation in helping to determine the proper clinical pathway, or order of treatment.

When looking at facility management in healthcare the literature is very limited and does not directly relate to the defined problem. Within contextual awareness for facility management, ontology has not been used to such a wide extent. The most common applications of ontology in terms of contextual awareness within the operation and maintenance of the facility are for building monitoring. One effort for real-time building monitoring has been attempted through using an ontology linked to building sensors in a multi-agent software framework (Dibley et al. 2012). This ontology takes into account the dynamic environment to aid in decision support while operating the facility. Outside of facility management, within the construction realm of building data, Hwang and Liu (2010) looks at taking real-time data and connecting it to a Building Information Model (BIM) with embedded semantics for aiding in project control.

A more common use for ontology within facility management is for data exchanges and management. Ontology has been used to document textual data from construction documents into an IFC-based model for facility management (Caldas and Soibelman 2003), for managing facility data in connection with a semantic-web database (Shevers et al. 2007), and collecting building data within an IFC file format for collaboration and sharing through the facility lifecycle (Vanlande et al. 2008).

Even though no facility management ontologies specific to healthcare were identified, there are two formalized taxonomies being developed to document facility related patient safety events by linking them to their causes. These are Common Formats by the Association of Healthcare Research and Quality (AHRQ) (www.ahrq.gov) and the International Classification for Patient Safety (ICPS) by the World Health Organization (WHO) (www.who.int). As discussed in Lucas (et al. 2011), these incorporate historical information and data about facility related events and have the potential to be used to develop response best practices or procedures. When implemented, Common Formats and ICPS can serve as a basis for a decision support system in responding to similar activities.

With these relevant research and application efforts, ontology and real-time data monitoring are more widely used in healthcare facilities to manage clinical data, especially pertaining to patient condition, than they are used to manage facility data. Within the facility management side, real-time data monitoring is more common for building systems such as temperature and humidity controls and is becoming more popular for automating these controls based on occupancy of space and other environmental factors. The clinical and facility management sides both use contextual reasoning in determining proper actions but a gap exists with the lack of information exchange between the two sides. The systems used by either side do not communicate needed information automatically. For instance, when a work order is created by facility management to do maintenance work in a patient room, the facility manager has no way of knowing if that room is occupied, nor can they block the room for maintenance to prevent clinical personnel from assigning the room within an electronic bed allocation system. To schedule the maintenance and conduct repairs without disrupting clinical operation requires coordination beyond what each party's electronic system can handle. More significantly, in an emergency situation, knowing what spaces are occupied and where patients are located in proximity to an emergency event is significant to the response. In this type of event, if a mechanical system has to be taken off-line, for example the oxygen system connected to an operating suite, it would be very important to know that when the system is shut down it is not being used in a critical operation by clinical personnel. Closing this data gap, with the aim of improving information exchange to enhance the operation and maintenance of healthcare facilities, is the goal of the proposed healthcare facility management ontology.

RESEARCH METHODOLOGY

The research discussed in this chapter leads to the creation of an ontological framework to support facility management in healthcare. This was done through a completion of multiple steps:

1. Documenting, in great detail, current procedures and process in response to emergency facility-related events within a healthcare environment.

2. Analyzing the mapped processes for individuals involved in response, systems used during the response, and information referenced during the response.
3. Documenting the communications needed with different individuals and systems.
4. Documenting the types of information needed during response and the source of that information (such as to an existing system, document, etc.).
5. Organizing the different information types within a product model to understand the conceptual hierarchy and relationship between the information types.
6. Develop an ontology that incorporated the relationships defined within the product model and links, queries, and filters information from the various identified systems.
7. Develop a conceptual model Graphic User Interface (GUI) to allow a user to interact with the ontology and retrieve necessary information during an event.
8. Use test cases of actual occurrences to ensure that the GUI and ontology can provide the needed information to the user during an actual event.

These steps are discussed in greater detail in the subsequent sections of this Chapter.

HEALTHCARE FACILITY MANAGEMENT ONTOLOGY

The healthcare facility management ontology is designed to help with decision support for facility managers through contextual reasoning. The ontology connects historical information and relevant real-time data from both clinical and facility management services within the healthcare environment. The historical information, building data (building information model and systems information), and regulatory/procedural information is all managed directly within the ontology. This information was organized into the domain taxonomy and represented as a product model to help in visualizing the relationship of the different information types. There are also lines of communication from the ontology to systems that provide the necessary real-time data. These include the building control systems, work order systems to track performed work, and clinical systems that provide patient data, service locations, and occupancy of spaces (Figure 3-2). The developed ontology is later linked to a series of Graphic User Interfaces (GUIs) to allow a potential user, facility management personnel, to access the needed information and allow for decisions support and reasoning in responding to defined events.

Methods

The ontology's development included basic steps of, (1) defining classes, (2) arranging the classes according to semantic and taxonomic relationships and (3) defining properties in each class. However there was a challenge in creating the ontology to

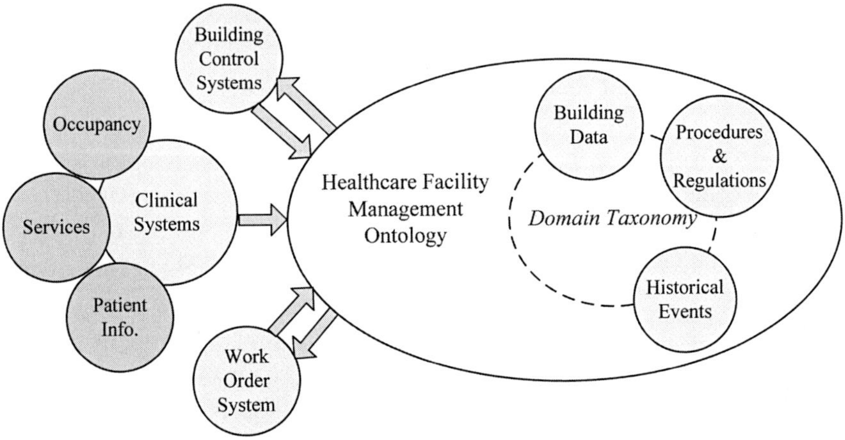

Figure 3-2. Ontology links and communications

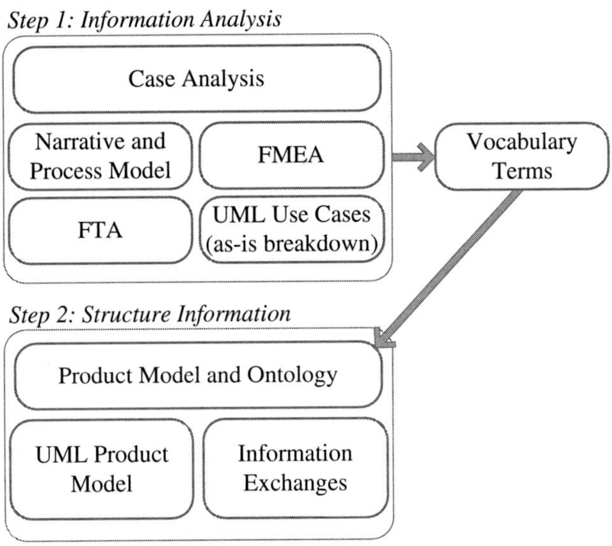

Figure 3-3. Research methodology steps

link two separate domains. This required defining overlaps in the separate knowledge of each domain and then structuring the ontology for both the healthcare and building industries, or more specifically for facility management within healthcare. This means each concept is defined in the same ontology. This required analysis of the information of each domain to determine how it was related.

Specific to this ontology, the development process is broken into two steps. They include information analysis and structuring the information so it's usable to support facility management response to patient safety events (Figure 3-3).

The information analysis was completed to determine the "vocabulary" from each domain. This was completed by looking at the overlap of data types involved in documented case studies. During the Information analysis, selected case studies were analyzed with a series of methods to define the "vocabulary" in the processes undertaken during the response to the event. These methods included developing a narrative and process model for each of the case-studies. The narrative and process model then served as a basis for conducting an Failure Mode and Effects Analysis (FMEA) and a Fault Tree Analysis (FTA) to identify possible failures and causes. Lastly, UML Use Cases were used to break down the processes, possible failures, and interactions between actors in a very detailed step-by-step analysis. The defined "vocabulary" later mapped to the ontology and covered the overlap of both domains, which is the result of the second step. In developing the ontology and structuring the information, the identified vocabulary was analyzed for its point of origin and original source. Historical and regulatory information sources were organized into a product model for easy management at first, to help in the understanding of information relationships and hierarchy, and then incorporated directly into the ontology. For real-time data sources, the systems of data origin were documented so they can later be connected to the ontology. The ontology serves as the information exchange and reasoning mechanism within the larger healthcare facility information management framework.

Information Analysis

The scenarios used as case studies were identified through a series of interviews with facility management and clinical personnel of a 500 bed university hospital in the United States that manages over 27,000 hospital admissions annually. In order to limit the scope of initial work for proof-of-concept, only scenarios involving facility management and mechanical systems were examined. The scenarios that were identified for further study had an impact or a significant potential impact on patient safety and clinical operations. The selected scenarios were documented and analyzed to determine proper vocabulary terms to use in developing the ontology (Lucas et al. 2013a). The analysis steps included:

1. Developing descriptive narratives of the case studies
2. Documenting as process models in Business Process Model Notation (BPMN)
3. Conducting Failure Modes and Effects Analysis (FMEA) and Fault Tree Analysis (FTA) to determine alternative model flows
4. Creating UML Use-cases to identify vocabulary terms

Descriptive narratives were created of the selected case studies that allowed for initial documentation and review by clinical and facility management personnel of the events. These allow for a general understanding of the steps that take place during the event's response and repair. A summary of a descriptive narrative is found below that will be used as example throughout the rest of this chapter.

Water Incursion in the Operating Suite

> The air-handling unit serving an Operating Room (OR) suite within the hospital malfunctioned. Water, from the chiller plant, was being pumped into the unit with a clogged pipe and water leaked from the unit. A nurse noticed the water coming from the ceiling within the operating suite over a corridor and OR and reported it to the Building Operation Center (BOC). The BOC is a call center for facility related issues. Upon investigation it is observed that water had damaged the ceiling, walls, and floor of the supply closet and adjacent hallway and OR and had also spread to the ceiling and floor two emergency room bays on the story below that housed the Emergency Department (ED). Care was taken by FM personnel, nursing staff, administration, and infection control to ensure that the situation was taken care of quickly and fully mitigated.

Since this case occurred during the weekend and early in the morning there were no operations currently underway so no patient was put in immediate danger. All scheduled surgeries in the operating suite had to be rearranged until the repairs were made. Overall expenses from this incident reached over $7 million in physical repair cost, replacement of damaged supplies cost, and lost revenue from the disrupted clinical services. It was noted during the case documentation phase that variations to this case, such as timing of the situation, proximity to patients, and location to certain services can cause major difference to the outcome and types of processes that would need to take place during the response. These alternatives were examined during the FMEA and FTA step of the case analysis.

Once the full descriptive narrative was created, a process model was developed. The process models were developed as BPMN diagrams. This allowed for documenting each individual step and decision in a high level of detail. The process model notation also allowed for allocating each step to a pool connected to an actor (group, individual, system). This allowed for easy visualization of who was responsible for each decision and action within the overall complex web of individual tasks that make up the event response. The process model also allowed for identifying which type of information and data systems were used or consulted during the process (Figure 3-4).

Since the completed process models showed the order of events as they happened with static variables, it was important to examine the effects of changing variables on the overall system. Since it was noted in earlier meetings with industry personnel that location, proximity to patients, and severity of the problem can all play a part in determining the proper response procedures, further analysis of the process models were conducted through a Failure Mode and Effects Analysis (FMEA) and Fault Tree Analysis (FTA).

The Failure Mode and Effects Analysis (FMEA) allowed for examining different variables and their effect on the failing systems including location of the failure and what part of the mechanical system failed. The FMEA documented each potential failure, cause of failure, effects of failure on the building system, effects on the health system and health safety, as well as different severity ratings

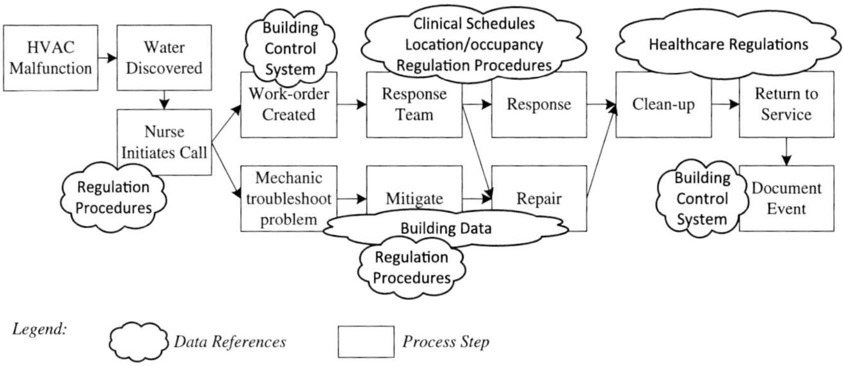

Figure 3-4. Example process model with data references

(Table 3-1). The FMEA showed how changing variables such as location, proximity to patients, and severity of failure within the mechanical system affect the threat to patient health and response to the situation. For example, if the situation had occurred in a janitor closet and not in a sterile supply closet, there would have been fewer damages as water does not have such a negative and costly effect on cleaning chemicals as it does on sterile medical instruments. Also, if a water stain, instead of standing water, was found within a corridor of an operating suite it would require immediate action but not such a dramatic chain of events as standing water because the damage of the water has yet to spread and does not pose as large of a threat. The FMEA allowed for documenting these variables, which in turn help to influence the design of the healthcare facility information management ontology.

The Fault Tree Analysis (FTA) takes into account each possible failure noted in the FMEA as a potential problem and identifies the immediate and root causes of each problem. FTAs show immediate causes back to a root cause of each identified problem. For instance, if there is water coming through the ceiling in the central sterile storage, one of the possible causes is that the air handler above the ceiling malfunctioned and is leaking water (Figure 3-5). Possible cause can be the air handler coil is backed up or a chiller line burst. For each of the possible causes there may be multiple levels of possible causes or a single root cause. The purpose of the FTA is to be able to track each problem to its ultimate root cause. This helps to support trouble shooting system problems by symptom within in the developed ontology.

The final step of the case analysis with the aim of identifying a list of vocabulary terms to be used within the ontology and data model was to develop a series of Unified Modeling Language (UML) Use-cases. UML Use-cases allow for showing a step-by-step process of one actor within a system. This actor can be computer program, individual, or group. Each use-case documents the specific interactions of one user during a single activity. Communications with other actors and data referenced or used during the activity are also noted within the

Table 3-1. Partial example of FMEA table

Failure Modes and Effects Analysis (FMEA)—Patient Safety										
Mechanical System over Operating Room Suite										
Item Number	*Item*	*Potential Failure Mode*	*Failure Cause*	*Facility Failure Effects*	*Health Failure effects*	*Likelihood of Occurance*	*Detection Method*	*Likelihood of Detection*	*Severity*	*Actions to Reduce Occurrence of Failure*
100.01	HVAC Airhandler	Chiller supply line leak over Janitor closet	leak at fitting/ oxidation and pin hole leak form	water in ceiling material of closet/fix and replace ceiling	Airborne contaminates and mold/ mildew	Medium	flow sensor or visual water mark on ceiling	Medium	Low	Regular maintenance of systems
100.02		Chiller supply line leak over non-sterile cooridor	Leak at fitting/ oxidation and pin hole leak form	water in ceiling material needs to be dried/ replaced	Airborne contaminates and mold/ mildew—Responsory problems to patients or other infections	Medium	flow sensor or visual water mark on ceiling	High	Medium	Regular maintenance of systems and training to report cases as soon as something is noticed
100.03		Chiller supply line leak over sterile supply	leak at fitting/ oxidation and pin hole leak form	water in ceiling, supplies need to be moved—replaced	Airborne contaminates and mold/ mildew in ceiling—possible contamination of supplies	Medium	flow sensor or visual water mark on ceiling	Medium	High	Regular maintenance of systems and training to report cases as soon as something is noticed

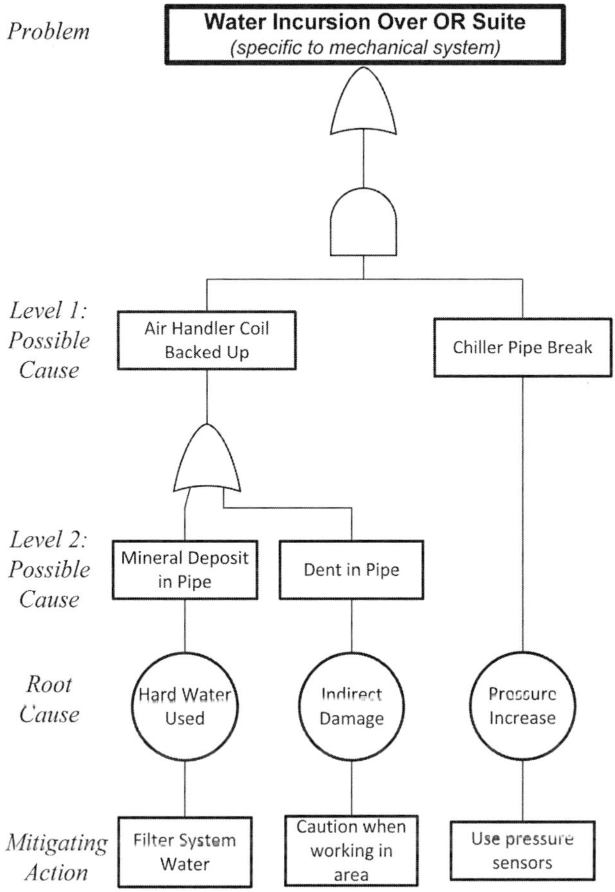

Figure 3-5. Partial Fault Tree Analysis (FTA)

use-cases. The use-cases allowed for identification of common terms within the domain that was used in responding to the event.

The use-cases were designed to include a base flow and alternative flows. The base flow documented the original narratives while the FMEAs and FTAs were used to design the alternative flows. Since one use-case describes one task, there are many use-cases for each case study. The use-cases fell into one of the major functions of the developed ontology. These functions are identified as *Initiate Event, Response, Diagnosis, Repair,* and *Recovery* as shown in Table 3-2.

Associated with each of the use-cases is Use-Case Diagram which visually connects different use-cases and different actors involved in the process. (Actors can be a person or system involved in the process.) An example of how a use-case is written is included below with the "Receive Water Incursion Call" Use-Case and its Use-Case Diagram in Figure 3-6.

Table 3-2. Use-Case Descriptions

Use-Case Type	*Case Analysis #1*	*Case Analysis #2*
Initiate Event	Water Incursion Reported	Temperature Complaint
	Water on Ceiling (Alternate)	Create Work Order
	Check Alarm I (Alternate)	
Response	Investigate Watermark (Alternate)	Shut Down Systems
	Maintenance Mechanic Response	Initiate Group #1 Response
	Contact Clinical Administration	Administrative Response
	Level I Response and Evaluation	
	Level II Response and Evaluation	
Diagnose	Check Alarm II	Check on HVAC
	Lookup Systems Information	
Repair	Schedule and Repair	Repair unit
	Work Scheduling	Contractor Repair
	Make Repairs to Unit	
	Repair Walls, Ceiling, and Floor	
Recovery	Put Systems back online	Put systems back online
	Initial Cleaning	Initial Cleanup
	Air Quality and Moisture Tests	Air Quality and Moisture Tests
	Review of Tests	Final Clean
	Final Cleaning	Reopen Building

Use-Case–2. Water on Ceiling

1. **Brief Description**
   This use case describes the reporting of the operating room nurse of a water mark on the ceiling as opposed to standing water, in the sterile supply room. The required responses are different.
2. **Use-Case Diagram**
3. **Preconditions**
   - The BOC Operator and Operating Room Nurse must have access to a phone.
   - The BOC Operator must have access to the Building Control System and it be online.

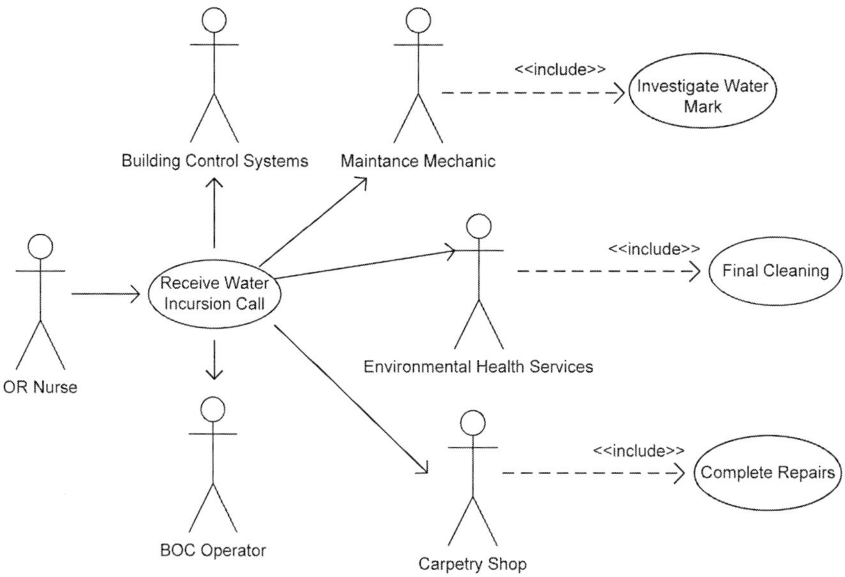

Figure 3-6. Receive Water Incursion Call Use-Case Diagram

4. **Basic Flow**
   - The BOC Operator receives a call from the Operating Room Nurse that there is a water mark appearing on the ceiling tile in the sterile supply room.
   - The BOC Operator creates a work order number from Building Control System and takes basic information from the nurse pertaining to the room, location of the leak and any other damages that are noticed.
   - The BOC Operator logs the call in the BOC Log Book and immediately pages the Maintenance Mechanic.
   - Include use case ***Investigate Water Mark***.
   - The BOC Operator waits for a report back the Maintenance Mechanic.
   - The BOC Operator receives a call from the Maintenance Mechanic with an update.
   - The BOC Operator reports in the Work Order that the problem was an overflowing evaporation tray and the mechanic was able to empty the pan and clean out the condensation drain which was blocked with dust build-up.
   - The BOC Operator dispatches the Carpentry Shop to replace the damaged tiles.
   - Include use case ***Complete Repairs.***
   - The BOC Operator dispatches Environmental Health Services (EHS) to the scene to evaluate and do a ***Final Cleaning.***

Table 3-3. Domain Vocabulary Terms used for generalization analysis in taxonomy structure

Reported Problem	*Identified Hazards*	*Damages*
Troubleshooting Source	Clinical Schedule	Services Affected
Complaint	Location of Complaint	Reporting Person
Dispatch Personnel	Response Protocol	Emergency Operation Plan
Expected Hazards	Extent of Damages	Source
Floor	Space	Room
Mechanical Zoning	Work Order	Shut-down Procedure
Mechanical Unit ID	Valve ID	Valve Location
Risk/Damage Level	Risk/Damage Incident Level	Occupancy
Space Use	Repair Estimate	Repair Schedule
Air Quality Testing	Contractor Repairs	Needed Work
Problem Location	Effected System	Available Supplies
Replacement Part #	Supplier	Problem Mitigation

By analyzing the use-cases, a list of vocabulary, or common terms, for the domain was identified (Table 3-3).

FRAMEWORK SCOPE AND DEVELOPMENT

Once the cases were analyzed and the list of vocabulary terms identified, the framework of the ontology was able to be put together. This required a structuring of the information types and some further analysis to incorporate all information exchanges between actors (Lucas et al. 2013b). The defined methodology for developing the ontology and making the necessary connections to other systems was broken down into the following actions:

- Analyze vocabulary for generalizations (generalized terms) — generalized terms are used in developing the taxonomy
- Define user support requirements (competency questions)
- Structure generalized terms into a taxonomy that shows relationships and the hierarchy of different types of information.
- Document systems interactions and information exchanges

The scope of the ontology and overall framework is to aid facility managers in responding to patient safety events. The vocabulary terms were first generalized based on the scope of the ontology as to remove any excessive redundancy and allow for inclusion of future event and problem types. The generalizations were tested later in the process with test cases. In order to stay within the scope, a series of main functions that the ontology needs to help assist with and provide semantic reasoning were identified. These included:

1. Identifying the problem
2. Determining the proper response and following through with it
3. Diagnosing the source of the problem
4. Determining and conducting repairs
5. Recovery of services
6. Documentation of the process and any facility changes

In order to keep the development within the defined scope and assisting within these areas, competency questions were developed that describe user support requirements of the ontology (Table 3-4). These competency questions list tasks for each of identified functions that the ontology would need to be able to complete.

Table 3-4. User support requirements

Framework Function	*Questions*
Identifying problem	What can cause the reported symptoms of the event?
	What is the source?
	Has the problem occurred before and what were the causes?
	What are the possible causes?
	We have health threats identified, what might be problem source?
Response	What response is required for the identified problem?
	Is there a related protocol?
	How do we mitigate the situation?
Diagnosis	What is the source of the problem?
Repair	Who is qualified to make the repair?
	What are the standards associated with the repair?
	What parts are needed and where can we get them?
	What are the procedures (drawings, specifications) for doing the work?
Recovery	What procedures are needed to get everything back online?
	What regulations do we follow to get to "business as usual"?
Documentation	How was the problem remedied?
	What work was completed?

ONTOLOGY FOR HEALTHCARE FACILITY MANAGEMENT

As described earlier in the chapter, the Healthcare Facility Management Ontology connects several different systems. These include real-time information and static information storage. The static information storage includes building product model data, procedures and regulation information, and historical events. This static information is stored within the ontology that is designed according to the user requirements listed above. In order to fully understand how all of this information relates, it was first developed as a product model. This product model helps to clarify the different data types and concepts and how they relate to one another. The other connections that the ontology is designed to handle are dynamic in nature. These are links to clinical systems, work orders, and building control systems. These dynamic links are the sources of real-time information and are the main variables as to how a response is handled.

Domain Taxonomy Organization

The static information was organized into a hierarchy of classes with defined relationships as defined in Lucas et al. (2013b) and shown in Figure 3-7 before it was incorporated into the ontology. The core of the hierarchy is the "Event" class. It is within the "Event" that the dynamic information is retrieved to start the response product by taking into account the problem type, location, proximity to patients, and identified hazards. From here the support information within the data model can help with the trouble shooting and response once it is incorporated into the ontological framework.

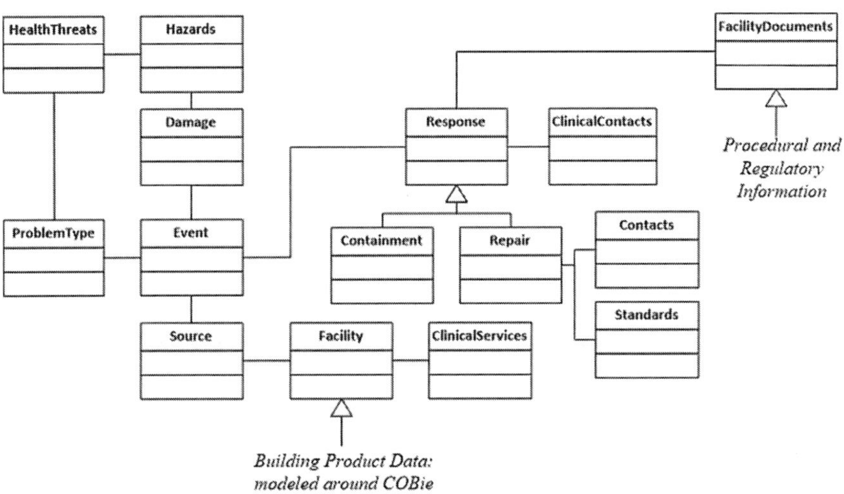

Figure 3-7. Domain Taxonomy Data Model—hierarchy and relationships of static information

The remaining classes within the data model are grouped together based on function. The "Response" class holds information to aid in determining and conducting the proper response. As two parts of the response, any responder (generally facility maintenance mechanics) will need to first contain the problem and then conduct the repair. The subclasses under the "Response" class are separated based on the distinct functional differences.

Also within the data model, the location of the event is represented within the "Facility" class. This class is formatted based on how the Whole Building Design Guide, Construction Operations Building Information Exchange (COBie) is organized. The reason for choosing COBie is because it targets the use of storing facility data throughout the facility's lifecycle. It offers a holding place for the building systems' data that can be accessed by the rest of the ontology when needed to help determine location-based information within the facility. This location-based structuring of systems information is important so the facility personnel conducting the response can know what system, or systems, is located adjacent to a space that can cause the identified problem. For instance, if there is water dripping from the ceiling in the operating suite, in order to proceed with the response, the mechanic can determine what systems, if they fail, are above that space and can exhibit the same symptoms.

The last cluster of classes revolves around the "HealthThreat" class and deal with managing the health threats that arise from a situation. Health threats can be determined using two different criteria. They can either be inherent based upon the problem type or by the hazard that is present. The hazards are typically associated with a damage that was caused during the problem. These classes hold the relevant data to assist in response and ensuring all possible health threats are taken into account.

INFORMATION EXCHANGES AND SEMANTIC REASONING

During the response process, the ontology uses the data from the real-time systems as well as other support information organized within the data model to support the user with reasoning in determining the proper course of action. The ontology is designed to aid the user throughout the entire process from the time a problem is reported and a work-order is opened through response, repair, and cleanup to get all systems operational.

An example showing the reasoning incorporated into the ontology is shown in Figure 3-8 as a sequence diagram. The sequence diagram comes part way through the response process where the ontology is aiding in determining who needs to be notified to participate in the event.

As shown in the sequence diagram, an operation "getClinicalContact" is called to list all contacts that need to participate in the response. This list is determined by numerous variables including the type of problem, what clinical

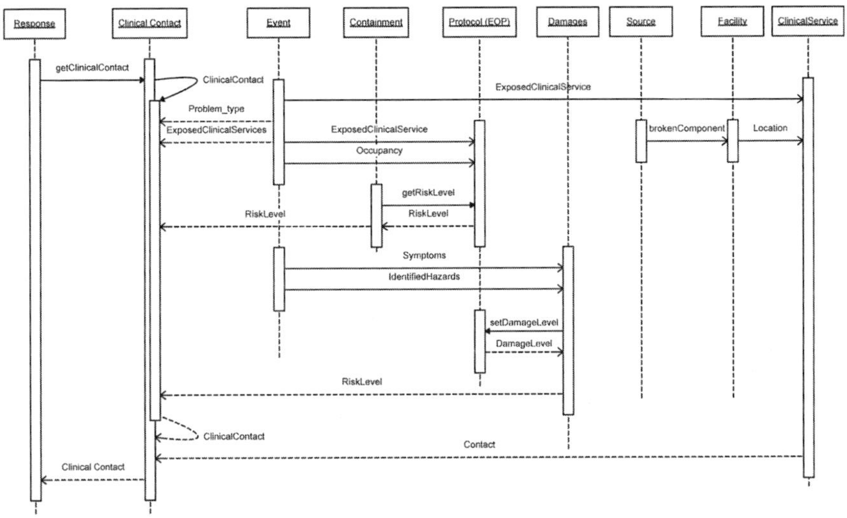

Figure 3-8. Sequence diagram sample—Identifying clinical contacts

services were exposed, what risk level the event is registered at, and the type of damages that have occurred. As shown in the sequence diagram. The operation is called. Then based on the type of problem, exposed clinical services, and occupancy (or proximity to patients and patient treatment areas) a risk level is determined. This risk level is defined through clinical protocol that is stored within the product model. Another operation to identify the hazards is also launched. This operation is based on the symptoms that were reported. This, in combination with the risk level, can then determine the damage level. With the defined risk and damage levels, it can be determined which clinical contacts make up the first part of the response team. This is completed through a developed matrix for emergency operations available in the clinical protocol manual. In order to have the entire response team, however, representatives from clinical areas that were exposed to the damage or hazard also need to be involved. This is completed through a third operation to identify the location where the hazards were reported, cross referencing it with who occupies the space, and then determining the proper point of contact for each clinical group affected. Once these three operations are complete, a list of clinical contacts necessary for the response can be compiled.

Some other functions that the ontology helps the user with are defined in Table 3-5.

With the ontology organized around these functions a proof-of-concept prototype and conceptual model was organized. These were used to do some preliminary testing to ensure that the ontology can provide the correct information for the identified functions and competency questions.

Table 3-5. Ontology reasoning

Function	*Variables Used*
Record Event	Work order creation Location (user input) Problem type and symptoms identified (user input)
Determine Response	Problem type, symptoms, location, patient occupancy
Determine Cause	Symptoms, location, facility system
Mitigation Procedures	Problem type, facility system
Documenting Damages	Facility system, identified damages (user input)
Mitigation of Hazards	Identified hazards (user input), damages, location
Identifying Health Threats	Location, occupancy, hazards
Scheduling/Documenting Repairs	Damages, cause

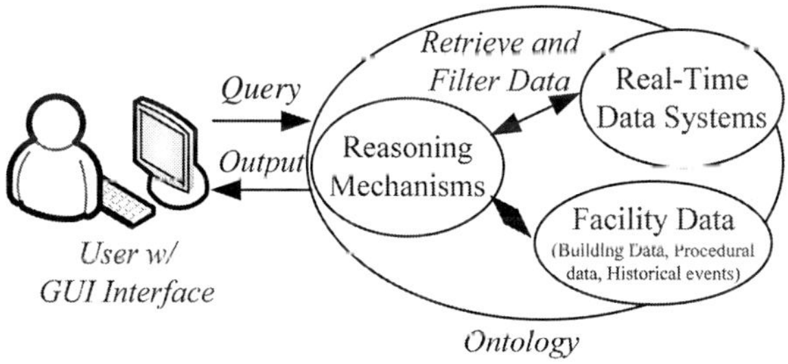

Figure 3-9. Conceptual model system architecture

CONCEPTUAL MODEL SYSTEM ARCHITECTURE

The conceptual model is depicted with the system architecture shown in Figure 3-9 and demonstrates how the user can interact with the ontology and the rest of the framework.

The user, through the designed Graphical User Interfaces (GUIs), places a query into the system. Based upon the documented conditions and the real-time status of relevant systems, the ontology, through its integrated reasoning mechanisms, determines the proper output to send back to the GUI and to show the user. This leads to the next step of the process where the user would make a necessary decision or complete a necessary action and use a subsequent GUI to aid in the task. The conceptual model is made up of multiple GUIs as defined in Lucas et al. (2013c). The conceptual model went through a development cycle as followed:

1. *Use case development:* Use cases were developed to detail interactions between the GUI, the product model, and other systems. They demonstrate specific operations of the ontology that the user can access through the GUI. The use cases helped to ensure that the GUIs were developed within the defined scope and intent of the ontology and connected framework.
2. *GUI Mapping:* Using the use cases as a basis, the seven main GUIs that are included within the conceptual model were mapped. The mapping identified interactions between the GUI, the ontology, the product model, and other systems. It also allowed for visually connecting background operations that the ontology was performing in sequence to the GUI interactions. A partial GUI interaction table is shown in Figure 3-10 that walks through the "Identify Clinical Contacts" operation discussed in Figure 3-8.
3. *Paper Prototyping:* Paper prototyping allows for using sketches to map out a system without the need for programming. This allows for a visual representation of GUI function without the need for total implementation.

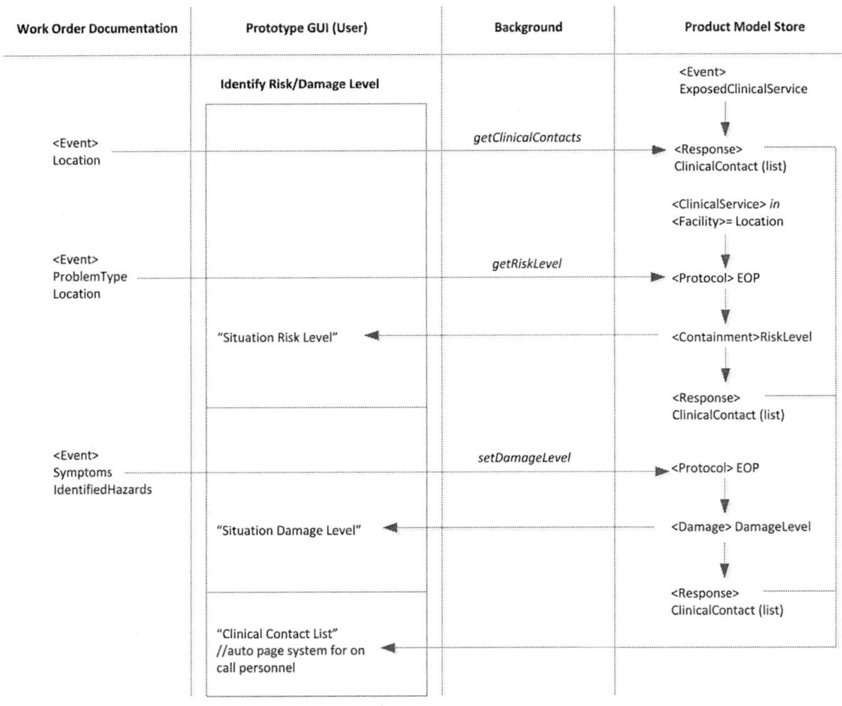

Figure 3-10. Partial GUI interaction table

4. *Conceptual Model:* A conceptual model of the GUIs was developed using Eclipse. Functions were then mapped to the GUIs and diagrammed. Classes within the product model and information exchanges from the ontology were noted within the conceptual model to demonstrate how the information exchange and interaction of the user with the information is handled.
5. *Test-case validation:* The test-case validation used two additional case studies to test the flexibility of the conceptual model's design. Test-case information was mapped on the developed GUIs to check that all information needed in the additional cases could be handled by the designed conceptual model and ontology.

GUI Development

A paper prototyping process was used to develop the GUIs. The paper prototyping allows for sketching user interfaces and quickly organizing how users are able to interact with the system. It was determined through the process that seven main GUIs were needed in order to fulfill the needs of the ontology and its defined scope and relating competency questions. These seven GUIs and a description of their function are listed in Table 3-6.

Each of the GUIs is designed around the same format (Figure 3-11). A user input/output menu on the left side of the screen with a model/document viewer on the right side. A menu and button system was chosen for ease of user interaction. The systems that facility maintenance personnel, the targeted user, are familiar with are mostly developed around a menu and button system for user input. The main interface was organized in this way to hopefully increase usability and ease of technology adoption with the target user.

Conceptual Model

The conceptual model was organized around the developed GUIs and mapped the functions of the GUI to the ontology and product model. This mapping allowed for a visual identification of where all information was entered into the overall framework and what operations took place in what order. The purpose of the conceptual model was to demonstrate the functionality of the ontology and how the user can retrieve needed information with its use.

One example of a demonstrated function within the conceptual model linked to the ontology is identifying the personnel contact information that make up the necessary response team. This function, as described earlier in Figure 3-8, is combined with the "Event Information Risk Level" and the "Event Damage Level". Both of which are defined based on location, problem type, and severity information that has been identified by an earlier operation. The GUI for defining the "Personnel Contacts" is shown in Figure 3-12.

Within this GUI, the information from the "Event" class that has already been defined is used to define the "Event Infection Risk Level" and "Event Damage Level". Based upon these two variables and the area that has been exposed to the

Table 3-6. Developed GUIs and their functions

GUI	*Function*
Event Information	• Classifies type of event • Determines required next step based on problem/location
Hazard Mitigation	• Determine mitigation steps associated with event hazards
Locate Source	• Output possible sources of problem based on location/problem and symptoms
Identify Risk and Damage Levels	• Determine Risk and Damage levels based on location/problem and symptoms • Determine proper response team that needs to be notified
Damages	• List damages (parts for repair) based on problem source • Document other damages (e.g. walls, floors, etc.) that are a result of the problem
Hazards and Health Threats	• Document additional hazards beyond those in the initial "Hazard Mitigation" • Determine health threats associated with hazards • Determine proper mitigation for hazards
Repairs	• Determine work to be completed • Determine party responsible for repair • Determine and allocate supplies needed

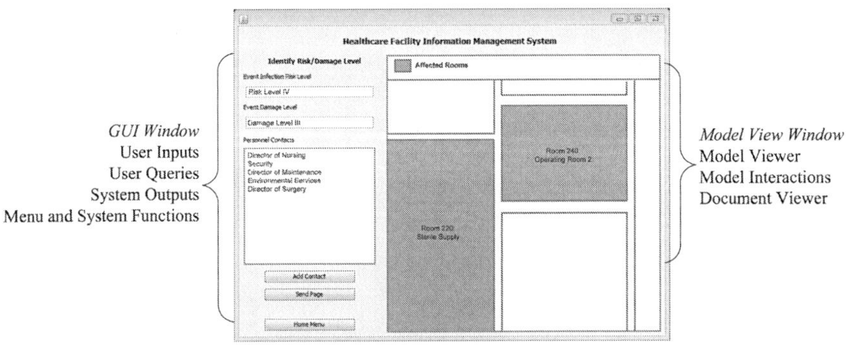

Figure 3-11. Design and format of GUI

problem, it can be determined which personnel or personnel groups need to be contacted to participate as members of the event response team. Each function within the conceptual model was mapped to ensure that adequate information is available through the ontology and framework to support response activities.

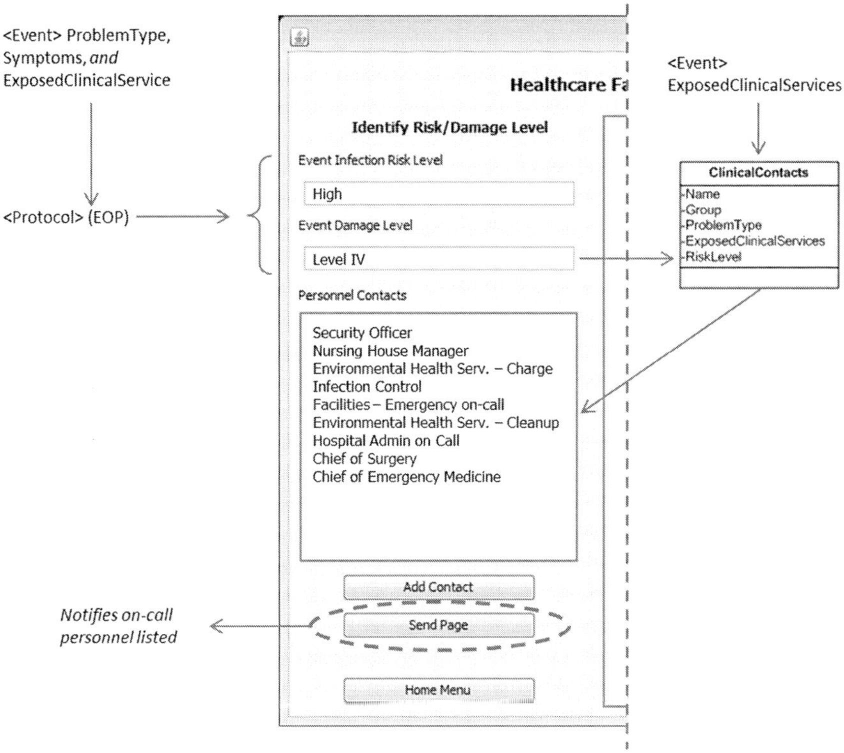

Figure 3-12. "Identify Risk/Damage Level" and "Personnel Contact" GUI

Test-Case Validation

The ontology and framework were designed around specific scenarios. In order to test that the framework held or is connected to the proper data and to ensure that the data classes were properly generalized and that the GUIs allowed the user to access the needed information, the ontology and framework need to be validated. This validation was completed using test cases.

Two test cases were identified to evaluate and validate the design of the conceptual model. These test cases allowed for ensuring that the designed functioned within the conceptual model worked beyond the scope of a case studies used during the conceptual model design. The test cases were documented in a similar fashion as the original case studies. Process models were created to document the information referenced, communications between individuals, and systems referenced during the event response. The process models helped to identify the exact pieces of information that were used during the response.

Once the information was identified for both of the test cases, the information was then populated within the ontology and corresponding GUIs for each test case. The GUIs were then used to replicate the actions identified in the process models for the test cases. In doing do, an expected result of information or available communications was compared to what the GUI and ontology allowed for representing. The expected versus available information and communications were compared to identify gaps within the framework. Appropriate modifications were then completed to fill the gaps and the original design cases were then re-examined to make sure the changes did not affect the ability of the framework to work with them. This test case validation process helps to make sure that the framework is generalized enough to work with a range of cases rather than only being able to work with the cases it was originally designed around.

RESULTS AND DISCUSSION

With preliminary feedback from the industry during industry meetings and review there was interest in the development of a system that would allow their personnel to more efficiently and effectively access needed information. It was felt from the industry that a tool that provided these functions might be able to allow their limited resources and personnel to more efficiently do their job to free up time to more consistently do preventive maintenance and other less important tasks that often are left incomplete.

The testing and validation that has been complete to point was done to quantitatively state if the framework (the ontology and integrated prototype) allows for structuring and retrieving of appropriate information. The types of information stored, the relationships of different classes of information, and the connection to other information systems was examined and tested through the conceptual model to make sure it can input proper information with logical connections. Future testing will require additional implementation of the prototype before it can be taken to industry professionals, who are the intended user, for review and completion of usability studies. Usability studies will test the actual acceptance and eventually the effectiveness of the ontology and framework.

Limitations of this research include its inclusion of only mechanical system issues. For the time being and the purpose of the completed research to demonstrate a proof-of-concept the inclusion of a single system type allowed for a very thorough analysis of response processes and information types needed. It is believed that other systems can be included within the developed framework and very easily implemented into the ontology as response to situations are defined by protocol that are similar no matter what the ultimate cause of the problem is. Future directions of research will include further implementation of the developed system for testing.

ACKNOWLEDGEMENTS

The background for this research is from a project funded under grant number 1 R02 HS19074-01 from the Agency for Healthcare Research and Quality (AHRQ),

U.S. Department of Health and Human Services (HHS). The findings and conclusions in this document are those of the authors, who are responsible for its content, and do not necessarily represent the views of AHRQ. No statement in this manuscript should be construed as an official position of AHRQ or of the U.S. Department of Health and Human Services.

References

Bakken, S., Cimino, J. J., and Hripcsak, G. (2004). "Promoting Patient Safety and Enabling Evidence-Based Practice through Informatics." *Medical Care*, 42(2, Supplement: Measuring and Improving Health Care Quality), II49–II56.

Beyan, O. D. and Baykal, N. (2012). "A Knowledge Based Search Tool for Performance Measures in Health Care Systems." *Journal of Medical Systems*, 36(2012), 201–221.

Caldas, C. H. and Soibelman, L. (2003). "Integration of Construction Documents in IFC Project Models." American Society of Civil Engineers, 1–8.

Dibley, M., Li, H., Rezgui, Y., and Miles, J. (2012). "An ontology framework for intelligent sensor-based building monitoring." *Automation in Construction*, 28(2012), 1–14.

Dijkstra, K., Pieterse, M., and Pruyn, A. (2006). "Physical environmental stimuli that turn healthcare facilities into healing environments through psychologically mediated effects: systematic review." *Journal of Advanced Nursing*, 56(2), 166–181.

Fenza, G., Furno, D., and Loia, V. (2012). "Hybrid approach for context-aware service discovery in healthcare domain." *Journal of Computer and System Sciences*, 78(2012), 1232–1247.

Hu, Z., Li, J. S., Zhou, T. S., Yu, H. Y., Suzuki, M., and Araki, K. (2012). "Ontology-Based Clinical Pathways with Semantic Rules." *Journal of Medical Systems*, 36(4), 2203–2212.

Hwang, S. and Liu, L. Y. (2010). "BIM for Integration of Automated Real-Time Project Control Systems." American Society of Civil Engineers, 509–517.

Iftikhar, S., Nawaz, F., Ahmad, H. F., and Fatima, K. (2010). "Efficient discovery of OWL-S based HL7 compliant healthcare web services profiles." *Emerging Technologies (ICET), 2010 6th International Conference on*, 287–292.

Imam, F. T., MacCaull, W., and Kennedy, M. A. (2007). "Merging Healthcare Ontologies: Inconsistency Tolerance and Implementation Issues." *Computer-Based Medical Systems, 2007. CBMS '07. Twentieth IEEE International Symposium on*, 530–535.

Isern, D., Sanchez, D., and Moreno, A. (2012). "Ontology-driven execution of clinical guidelines." *Computer Methods and Programs in Biomedicine*, 107(2012), 122–139.

Lucas, J., Bulbul, T., and Thabet, W. (2013c). "A Pilot Model for A Proof of Concept Healthcare Facility Information Management Prototype." *ITCon*, In Press, Accepted for Publication.

Lucas, J., Bulbul, T., and Thabet, W. (2011). "A Lifecycle Framework for using BIM in Healthcare Facility Management." *Proceedings of the CIB W78-W102 2011: International Conference — Sophia Antipolis*, France, 26-28 October, http://itc.scix.net/cgi-bin/works/Show?w78-2011-Paper-73,.

Lucas, J., Bulbul, T., Thabet, W., and Anumba, C. (2013a). "Case Analysis to Identify Information Links between Facility Management and Healthcare Delivery Information in a Hospital Setting." *Journal of Architectural Engineering*, In Press: Accepted for Publication.

Lucas, J., Bulbul, T., Anumba, C. J., and Messner, J. (2011). "Evaluating the Role of Healthcare Facility Information on Health Information Technology Initiatives from a Patient Safety Perspective." American Society of Civil Engineers, 720–727.

Lucas, J., Bulbul, T., and Thabet, W. (2013b). "An object-oriented model to support healthcare facility information management." *Autom. Constr.*, 31(2013), 281–291.

Mokkarala, P., Brixey, J., Johnson, T. R., Patel, V. L., Zhang, J., and Turkley, J. P. (2008). "Development of a Comprehensive Medical Error Ontology." *Advances in Patient Safety: New Directions and Alternative Approaches (Vol. 1: Assessment)*, Henriksen K., Battles J. B., Keys M.A., et.al., ed., Agency for Healthcare Research and Quality, Rocville, MD: Available from: http://www.ncbi.nlm.nih.gov/books/NBK43635.

Rajan, J. M. and Lakshmi, M. D. (2012). "Ontology-based Semantic Search Engine for Healthcare Services." *International Journal on Computer Science and Engineering (IJCSE)*, 4(04), 589–594.

Rector, A. L., Qamar, R., and Marley, T. (2009). "Binding ontologies and coding systems to electronic health records and messages." *Applied Ontology*, 4(1), 51–69.

Shevers, H., Mitchell, J., Akhurst, P., Marchant, D., Bull, S., McDonald, K., Drogemuller, R., and Linning, C. (2007). "Towards digital facility modelling for Sydney opera house using IFC and semantic web technology." *ITcon*, 12(2007), 347–362.

Vanlande, R., Nicolle, C., and Cruz, C. (2008). "IFC and building lifecycle management." *Automation in Construction*, 18(1), 70–78.

Zeshan, F. and Mohamad, R. (2012). "Medical Ontology in the Dynamic Healthcare Environment." *Procedia Computer Science*, 10(2012), 340–348.

CHAPTER 4

Ontology to Support Multi-Objective Integrated Analyses for Sustainable Construction: A Conceptual Framework

Y. Zhu*

Abstract: *A typical construction project involves many parties with different objectives. Sustainable construction requires multi-dimensional thinking of social, economic and environmental implications. Therefore, a meaningful analysis of sustainable design options depends on a large amount of economic, technical and environmental data from a life cycle perspective. Although a significant amount of effort has already been put into research and development of integrated solutions that address traditional project objectives such as time, cost, safety and quality, environmental impacts represent a new dimension and need to be integrated with other project objectives. This chapter discusses the need for an ontology-based solution to integrate environmental protection considerations into project performance assessment. An ontology representation of integrated time, cost and environmental impacts (TCEI), called TCEI-Ontology, is presented. Ideas of reusing concepts in the Industry Foundation Classes (IFCs) are also discussed. The potential application of TCEI-Ontology to solve integration issues in time, cost and environmental impact analyses is illustrated with a case study. Further development and tests are needed to determine the feasibility and the strengths and weaknesses of the proposed approach.*

INTRODUCTION

Sustainable construction refers to building and construction practices, in which the impact of buildings and their construction processes on the natural environment is carefully considered and addressed during the design and

*Associate Professor, OHL School of Construction, Florida International University, Miami, FL 33174; PH (305) 348-3517; FAX (305) 348-6255; email: zhuy@fiu.edu

construction of buildings. Often sustainable construction is considered as a mechanism to deal with the issue that buildings and their construction processes account for a large amount of energy consumption and carbon emissions throughout the life cycle of buildings. Examples of sustainable construction are abundant such as LEED certified buildings. On the other hand, building and construction projects always have multiple objectives. Traditional objectives such as cost, time and quality need to be satisfied in addition to new objectives such as environmental performance.

Design plays a significant role in sustainable construction because decisions that can impact the most on cost, installation time and environmental performance are made during a design phase. Recent research in building design indicates that the most efficient, best performing, and most environmentally sustainable buildings are developed utilizing integrated project delivery (IPD), in which relevant disciplines work together from project conception, design to construction of high performance buildings. Such an integrated team is able, from the outset of the project, to consider multiple building and construction objectives and constraints as an integral part of a design process, treat a building and its components as a whole system, and develop integrated solutions, leading to better engineered, more efficient, and lower cost buildings (Molenaar et al. 2009, Yudelson 2009). However, IPD is a complex process involving multiple parties with different interests and objectives and demanding stronger design phase support. Although design support tools have been around for long time, there is a lack of integrated tools to support systematic optimization of designs.

During design, decisions are often made individually to specific building components or systems without considering combined effects to a building as a whole. Thus, the design process of sustainable construction projects involves the proper selection of building products, materials and systems that can effectively form a building design to satisfy its overall social, economic and environmental objectives. One critical issue is the evaluation of the performance of building materials, products or systems, as well as the performance of a building. Often, performance metrics are designed to serve multi-objective evaluations but data to support such evaluations are typically fragmented as they are collected from different sources. The data fragmentation problem has long been observed and discussed in building and construction research communities (e.g., Zhu et al. 2006). Over the last decades, solutions such as the Industry Foundation Classes (IFCs) have been proposed (e.g., Mao et al. 2007). Environmental impacts, as a new dimension of project objectives, add to the need for data integration, forming another level of complexity to the data fragmentation problem (e.g., Marzouk et al. 2008, Ozcan-Deniz et al. 2011).

Ontology, as a solution to problems in data and system integration and knowledge management has been studied extensively in the construction industry. For example, Lima et al. (2005) discussed an ontology-based portal for knowledge management in the construction domain. Rezgui (2006) proposed using ontology, as the conceptualization of the domain, to "ensure relevance, accuracy and completeness of information." Pandit and Zhu (2007) discussed the application of ontology to support designs that involve with Engineer-To-Order (ETO) products. Anumba et al. (2008) discussed the use of ontology as a means to overcome interoperability problems in knowledge management in construction. While many of those studies have a clear focus on knowledge management and system integration, other researchers discussed reusing existing conceptual models and converting them into ontology (e.g., Schevers and Drogemuller 2005, Beetz et al. 2009).

Using ontology to integrate environmental impacts with traditional project performance criteria is a viable solution and becomes critical to building design and construction analyses. In the following, the author first introduces life cycle assessment in the context of supporting integrated analyses. The need for ontology to support the integration of environmental life cycle assessment in traditional building or construction performance analyses is elaborated by using a case study. The design of TCEI-Ontology is also discussed, followed by a demonstration case. However, it is not the intention of this study to develop full-scale life cycle assessment ontology. Rather, the purpose of this study is to demonstrate the potential of using ontology to solve data and process integration for building scale multi-objective analyses and knowledge representations. In the meantime, a high-level ontological structure for integrated analyses is presented.

LIFE CYCLE ASSESSMENT FOR EVALUATING ENVIRONMENTAL IMPACTS

Life cycle sustainability assessment provides a framework for quantitatively evaluating social, economic, and environmental impact of products and materials from a life cycle perspective (UNEP 2011). In this study, life cycle assessment refers to environmental life cycle assessment, or E-LCA, which has received a large amount of attention for decades. It provides a scientific foundation for green building assessment. The advantages of applying life cycle assessment in the building and construction sector are well recognized and documented (e.g., Sharrard 2007, Bilec et al. 2006). For this reason, an existing green building rating system, the Leadership in Energy and Environmental Design (LEED) has started its work to incorporate life cycle assessment into its rating methodology (e.g., LEED 2010).

For decades, research studies and applications of life cycle assessment in the construction industry have been reported in areas of evaluating building products and systems, and construction processes. Singh et al. (2011) classified life cycle assessment applications in building construction into five categories, construction product selection, construction systems and process evaluation, tools and databases, and methodological developments. Khasreen et al. (2009) found that existing studies could be in general classified into two groups, whole process construction versus building and material components. In addition, three types of life cycle assessment methods are commonly cited, including process-based (e.g., Lippiatt and Boyles 2001, Zhang et al. 2006), input-output analysis (e.g., Matthews and Small 2001) and hybrid (e.g., Suh et al. 2004).

Life cycle assessment at building level is unique, because a typical construction process involves many parties with different objectives. Even though reducing negative environmental impacts is among those objectives, environmental conscious designs are typically subject to other project constraints such as cost, quality and safety. Thus, an integrated analysis, depending on a large amount of economic, technical and environmental data from unit processes to the entire construction process with a life cycle perspective, is key to the success of LCA applications at building level. However, existing economic, environmental and technical data and related information models are typically developed independently without considerations to support integrated analyses. Thus, major challenges arise from insufficient semantic and contextual definitions of life cycle assessment data, and the lack of interoperability with other types of data.

Studies on formalizing life cycle assessment have recently been reported in Europe and Brazil. For example, Bräscher et al. (2007) discussed the need for and the methods of developing life cycle assessment ontology for Brazilian industries. Moreno et al (2011) discussed the role of product data technology in exchanging reliable life cycle assessment data and knowledge in Europe. While neither of the two papers contains any details of life cycle assessment ontology, an example of assessing an individual home is available at the website of the CAESAR Research Projects (http://www.caesarsystems.co.uk/research_projects/DEPUIS/reference_data_on_the_web/lca_for_an_individual_house.htm#lca_for_an_individual_house). In addition, the need for using ontological approaches to integrate life cycle assessment and applications such as product data management (e.g., Ostad-ahmad-Ghorabi et al. 2012, Moreno et al. 2011) was clear to the research community and reported in contemporary literature.

Apparently, life cycle assessment ontology involves with many concepts covering a wide range of areas including manufacturing, transportation, construction, occupancy and end-of-life, as well as different environmental impact categories. However, the design of life cycle assessment ontology usually does not aim at supporting integrated analyses, which leads to gaps between existing building and construction related ontologies and life cycle assessment ontology. Thus, having an underline conceptualization that integrates building construction concepts and life cycle assessment concepts can be useful. In the following, a case is used to illustrate the need and the conceptualization of the proposed ontology framework.

A MOTIVATING CASE

In a typical design process, designers often compare design options at building level, but details regarding components or material selections are not necessarily

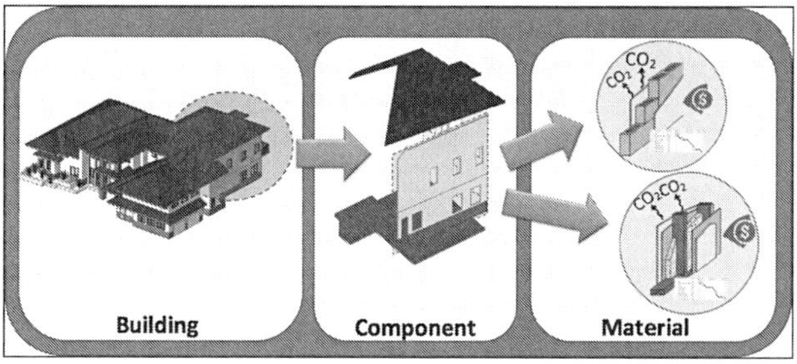

Figure 4-1. Optimization process

formalized for reasoning and supporting design selections at building level. Currently, reasoning processes are still mostly manual. While it may be debatable as to if such processes can be fully automated, information related to material or component properties, their functions, scope, constraints, and aggregation methods can be formalized to support designers in decision-making.

Figure 4-1 shows that the selection of residential building designs can be traced down to the selection of many building materials and components. For example, a building may use any of the three options for its exterior walls, structural insulated panels (SIPs), steel framing or wood framing. Each bears different implications to time, cost and environmental impacts. Similarly, there are many different material options that can be applied to other building components such as roofing, slabs, doors and windows, and interior walls. Thus, the final decision about a design is a combination of decisions over tens of thousands of building materials and components. When comparing different design decisions, it can be very helpful if rationale about decisions on certain materials and components can also be compared and reasoned.

Figure 4-2 further illustrates the complexity involved in comparing and evaluating designs with three options. In reality, the selection process can be rather subjective, thus having a mechanism to support reasoning is very important. At any given time, a reasonable designer makes a decision on the selection of a material based on a set of criteria. While it is not the objective of this study to develop an expert system to capture design intent, this study is focused on developing a framework of ontology that can help designers tracking the rationale of design selections at material, component, and building levels. In this way, designers can later determine why a certain material is selected based on the goal, scope and function that the material is intended to support.

There are two related issues in this case. First, when a designer compares the three options, how much confidence does a designer have about the scope of time, cost and environmental impact data associated with the three options? In other words, are the options comparable? What are the reasons that they are or are not comparable? Second, at a lower level such as materials or building components, how can one quickly know whether data often coming from different sources have the same or compatible scope? The first issue is related to the second one, which is

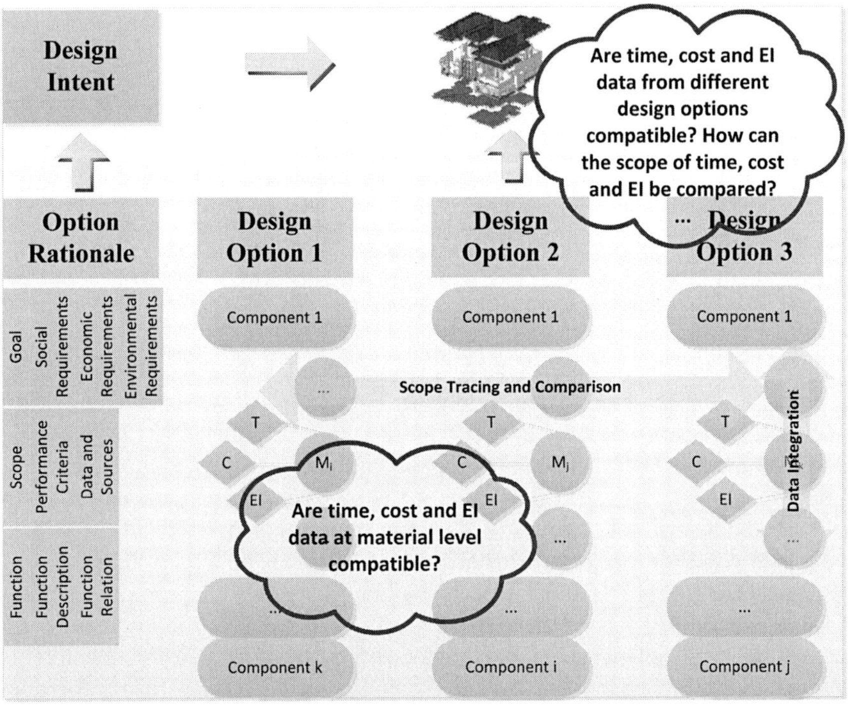

Figure 4-2. High level conceptual framework of design selection

more fundamental. Providing abilities for designers to track the rationale of their decisions at different levels is thus very important.

ONTOLOGY DEVELOPMENT

Ontology is a study about what exist in the world, or an explicit specification of conceptualization (Gruber 1993). Examples of applying ontology to the integration of software applications in the AEC industry are abundant. For example, Zhong et al. (2012) developed ontology-based semantic modeling of regulation constraints for automated construction quality compliance checking. Rezgui et al. (2011) provided a comprehensive review of applications of product data technology in the construction industry and promoted the use of ontology instead of data-centric approaches. Pandit and Zhu (2007) discussed an ontology-based approach to support decision-making for the design of engineer-to-order products in the construction industry. Many other similar studies were focused on sharing knowledge or integrating heterogeneous data, covering a wide range of fields in construction including cost (Staub-French et al. 2003 (a)) and scheduling (Staub-French et al. 2003 (b). In addition, the Industry Foundation Classes (IFCs) can provide a semantic-rich foundation for integrating or sharing cost and schedule data (Froese and Yu 1999). Therefore, in this chapter the author discusses ontology for environmental life cycle assessment with a focus on its integration with construction cost and time analyses.

There is a significant amount of literature on ontology development. The simple knowledge engineering methodology (Noy and McGuinness 2001) is used to develop the ontology framework, TCEI-Ontology, in this chapter. The methodology defines seven basic steps, including:

1. Determining the domain and scope of ontology,
2. Reusing existing ontology,
3. Enumerating important terms in the ontology,
4. Defining classes and class hierarchy,
5. Defining slots or properties of classes,
6. Defining facets of slots, and
7. Creating instances.

Concepts in TCEI-Ontology belong to two general domains, the construction management domain and the life cycle assessment domain. Many of those concepts are available in the Industry Foundation Classes (IFCs). In the following, LCA-related concepts are presented first, followed by other concepts.

LCA-Related Concepts

In this section, a list of typical environmental life cycle assessment concepts are generated, including those related to goal and scope definition, inventory analysis, and impact assessment.

Table 4-1 and Figure 4-3 show a list of major concepts associated with the goal and scope definition phase of LCA. The root concept in this segment is LCAStudy, which is defined by FunctionUnit, Goal, Project, Scope and TimeStamp. TimeStamp and FunctionalUnit can be defined by reusing IFC definitions. The project definition in an LCA study has broader meaning than the project definition in IFC, because ifcProject only covers design, engineering, construction, and maintenance activities and does not include other stages in the life cycle of a product. So, ifcProject is used to complement Project. The Goal concept determines the Scope of an LCA study. In addition, Goal is also defined by IntendedUser. Without providing too many details, the concepts are still high-level concepts. For example, the intended use is a major concept to define the goal of an LCA. It contains many different options such as supporting broad environmental assessment or establishing baseline information. Such details can be added later to enrich the definitions of concepts.

Major inventory analysis concepts are collected in Table 4-2 and modeled in Figure 4-4. These concepts are grouped around Process, Input and Output. Process refers to all sub-processes or unit processes that are included in an LCA, therefore its scope is larger than ifcProcess, which only includes design, engineering, construction, or maintenance activities. The Process in the tcei namespace

Table 4-1. Major LCA goal and scope concepts

Goal and Scope Definition		
Concept	*Description*	*IFC*
Functional Unit	Defines the performance characteristics of a product or process.	ifcUnit
Geographical area	Place of a project to which an LCA is performed.	ifcAddress
Goal	Defined by intended use, intended audience and LCA type	N/A
Intended audience	End users of an LCA study	ifcActorRole
Intended use	Support broad environmental assessment, establish baseline information for a process, rank the relative contribution of individual steps or processes, identify data gaps, support public policy, support product certification, and provide information and direction to decision-makers.	N/A
LCA type	Refers to a specific LCA or a generic LCA	N/A
Project	Refer to the actual product or process to be studied.	Partially available ifcProduct
Scope	The scope of an LCA study defined by function of product systems, functional unit, allocation procedure, data requirements, initial data quality requirements, and impact categories.	N/A
Time Stamp	Time when an LCA study is performed.	ifcCalenderDate

contains definitions that complement those in ifcProcess. In addition, a Process is associated with a life cycle stage, which is defined by an IFC definition, ifcEnvironmentIMpactCategoryEnum. This concept contains major life cycle stages such as extraction, manufacturing, installation, and disposal. It also allows for extensions. Each Process is also associated with LCIInventory, which determines the allocation method, cut off rules, and the inventory analysis method. Through InventoryList, inventory data are linked to each Process. The inventory data are described by data date and data location, which can be used later to trace differences in various design assessments.

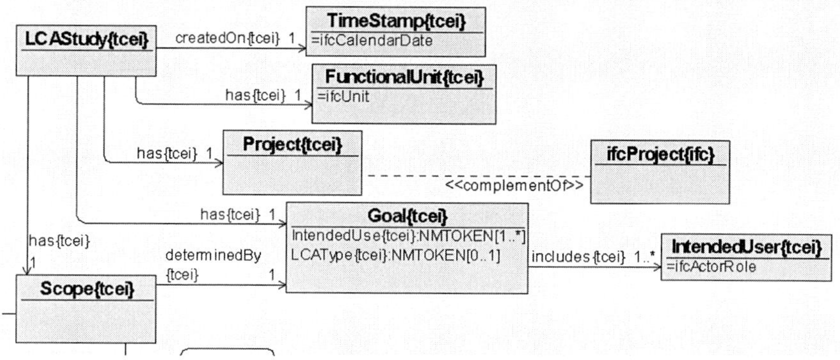

Figure 4-3. Goal and scope concepts

Input is composed of Energy, Water, ServiceGoods, Materials and ifcConstructionResource. They are disjoint and complementary to each other. The ifcConstructionResource is a super type of ifcConstructionEquipmentResource, ifcLaborResource, ifcCrewResource, ifcSubContractorResource, ifcContructionProductResource, and ifcConstructionMaterialResource. The ifcConstructionMaterialResource can be linked to ifcProduct, which represents products used by a construction project. In an LCA, there may be other types of intermediate materials that are not in the form for consumption by a construction project, thus the concept of Materials in the tcei namespace is included.

Output is consisted of ifcProduct and EnvironmentalBurden. ifcProduct can be a building component or a building itself and EnvironmentalBurden includes emissions to air, water and land. The environmental burden can be extended to include resource depletion and fossil fuel consumption.

Impact assessment includes fewer high-level concepts as shown in Table 4-3. Their relationships are shown in Figure 4-5.

The impact assessment is directly related to the EnvironmentalBurden as part of Output. EnvironmentalBurden includes emissions to air, water and land. In addition, the concept can be extended to include other types of waste, or energy and material flows. These burdens are related to ImpactCategory based on specific classification and characterization methods. The ImpactCategory is associated with ifcEnvironmentalImpactValue, which can be used to specify impact categories to be studied and their corresponding values.

The connection between goal and scope concepts and inventory analysis concepts is accomplished by two concepts, Scope and Process. Process is treated as part of a scope definition. Similarly, the connection between goal and scope concepts and impact assessment concepts is established by Scope and ImpactCategory, another part of a scope definition (Figure 4-6).

Many details are not included due to the conceptual nature of this work at the current stage, for example, different types of input and output in an inventory analysis. Such details can be included in the future. In addition, conceptually, some of those concepts are specific to the TCEI-Ontology, while others can be defined

Table 4-2. Major inventory analysis concepts

Inventory Analysis		
Concept	*Description*	*IFC*
Allocation	Allocation of environmental burden to processes based on 1) avoiding allocation, 2) mass or energy content, or 3) economic basis.	N/A
Cut-Off-Rules	Threshold for including an input or an output by 1) percentage of mass, 2) percentage of a flow to the total economic value, 3) percentage of a flow to the total environmental load.	N/A
Data Uncertainties	Data uncertainties associated with inventory data.	N/A
Environmental Burden	Include emissions to air, land and water.	N/A
Input	Include energy, land, materials, services, and water to a process/unit process.	Partially available ifcConstructionResource
Inventory List	LCI used in a study	N/A
Inventory Analysis Method	Method used in inventory analyses 1) process-based, 2) input and output, and 3) hybrid	N/A
Life Cycle Stages	Typically include raw material extraction, raw material processing, manufacturing, use and maintenance, and disposal or recycling.	ifcEnvironmentalImpact CategoryEnum
Output	Output of a process/unit process, including intended products, emissions and wastes.	N/A
Process	Include a list of sub-processes and unit processes, as well as sequences to show the production steps of a product.	Partially Available ifcProcess

Table 4-2. Major inventory analysis concepts (Continued)

	Inventory Analysis	
Concept	*Description*	*IFC*
System Boundary	Define what are included in an LCA.	N/A
Unit process	The basic processes in an LCA that inventory data are associated with.	ifcProcess

by other ontology. Since there isn't life cycle assessment ontology that is already available publically, it is hard to determine specifically what should be defined by TCEI-Ontology and what can be reused from other ontology at this moment. Therefore, the concepts in the life cycle assessment domain are mostly included in the TCEI-Ontology namespace, tcei.

Non-LCA-Related Concepts

Except for concepts related to LCA, there are concepts related to cost and time. Many concepts in the Industry Foundation Classes (IFCs) can be used to model non-LCA-related concepts. Table 4-4 shows the two most important concepts for time and cost representations. Figure 4-7 shows the two concepts in the overall model (Figure 4-8), where association models in IFCs, which represent dynamic yet complicated relations between concepts, are not used in TCEI-Ontology in order to keep this initial effort simple.

The ProjectCost concept, an equivalent concept to ifcCostItem, can be used to represent the cost of products, services, or life cycle cost. According to IFC (http://www.buildingsmart-tech.org/ifc/IFC2x3/TC1/html/index.htm), ifcCostItem can

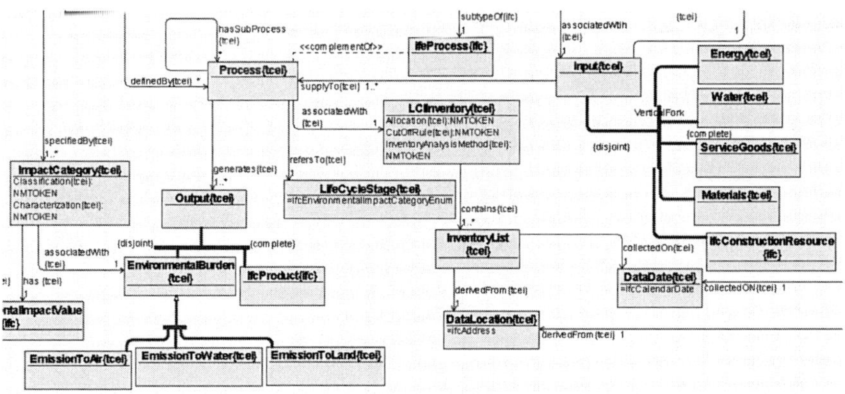

Figure 4-4. Inventory analysis concepts

Table 4-3. Major impact assessment concepts

Impact Assessment		
Concept	*Description*	*IFC*
Classification	Assign LCI to a particular impact category.	N/A
Characterization	Convert LCI results in the same category to a common unit based on characterization factors.	N/A
Grouping	Sort impact categories according to a particular characteristic such as geographic area or population.	N/A
Impact Category	Include typical impacts such as global warming, resource depletion, smog, human toxicity, ozone depletion, eutrophication, water use, land use, acidification and eco-toxicity.	Partially available ifcEnvironmental ImpactValue
Normalization	Represent impact results after characterization using a reference value.	N/A
Weighting	Assign a weight to each impact category.	N/A

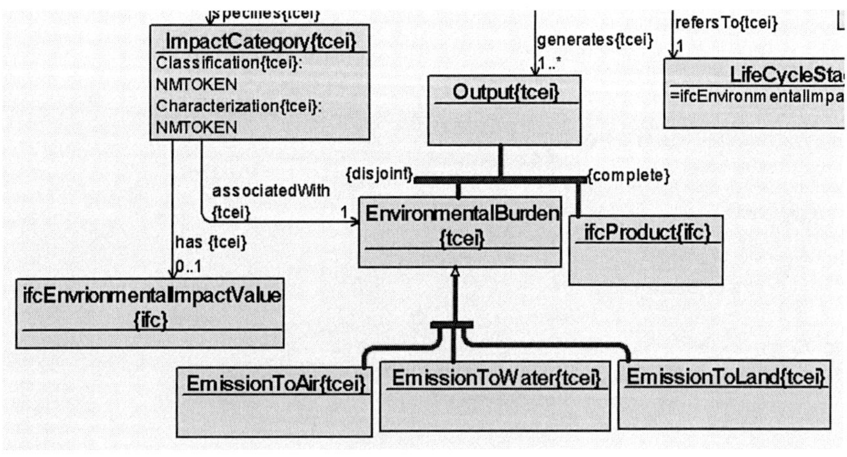

Figure 4-5. Impact assessment concepts

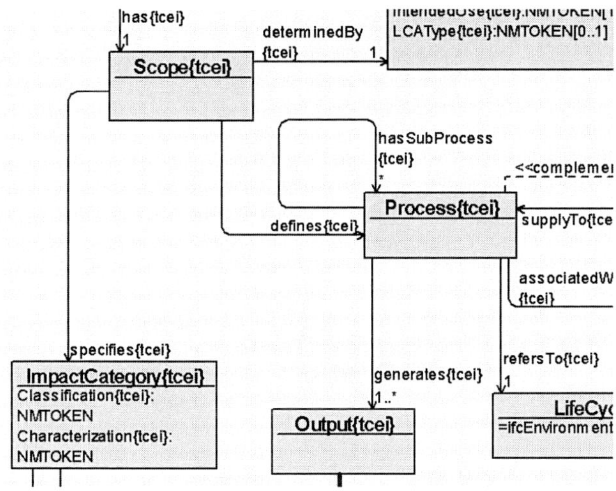

Figure 4-6. Connections between concept groups

have a nested structure to create a more complex cost component. The cost is determined by two other components, quantities and unit prices. Quantities are modeled by Input and unit prices are represented by ifcCostValue. On the other hand, the concept of time is modeled by ifcWorkPlan, which contains the start time and the finish time of a project. ifcWorkPlan can be associated with ifcTask, which is a subtype of ifcProcess. In this way, both time and cost concepts are linked with concepts of LCA.

Ontology Representation

In order to test the idea of using ontology to support integrated studies of time, cost and environmental impacts, ontology that captures major concepts of life cycle assessment, cost, and scheduling is modeled (See Figure 4-8). The diagram is developed by using a graphic ontology modeling software tool, called OWLGrEd, developed by the Institute of Mathematics and Computer Science at the University of Latvia.

Table 4-4. Non-LCA concepts

Non-LCA Concepts		
Concepts	*Description*	*IFC*
Project Cost	Cost of a project defined by cost items, their quantities and unit costs.	ifcCostItem, ifcCostValue
Project Time	Refers to the construction time of a project.	IfcWorkPlan, IfcTask

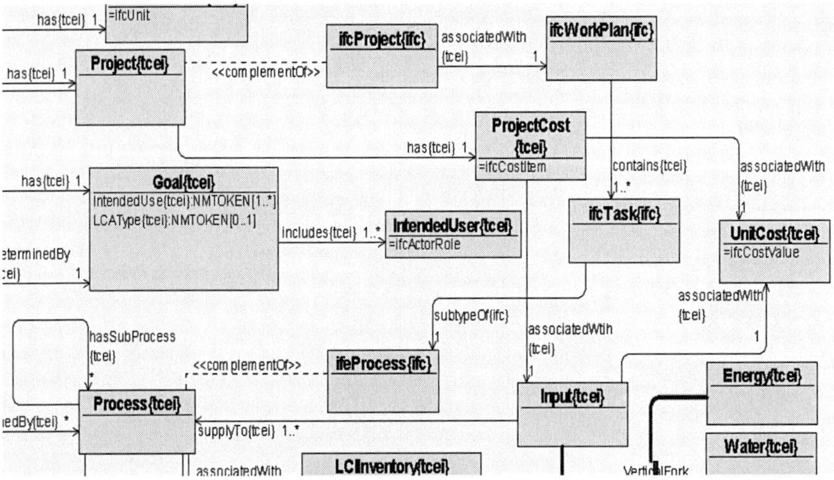

Figure 4-7. Time and cost concepts

In this ontology, the connecting point between time, cost and environmental impacts is the concept of Input, which can represent resources used in construction processes. From the perspective of life cycle assessment, Input is part of what goes into a unit process. Input is equivalent to the concept of construction resources, when a unit process belongs to the construction stage of life cycle assessment. This logical connecting point allows performing analyses of construction cost, construction duration and environmental impacts in an effective way. Input includes materials, energy, water, and service goods, as well as construction resources such as labor, equipment and construction products.

Lifecyclestage associated with Process sets a condition, in which only the construction phase input is included in time analyses, because life cycle assessment may include intermediate products or materials in manufacturing and cost may be life cycle cost instead of construction cost. By further connecting ifcProcess with ifcWorkPlan, the definition makes it clear that only construction phase resources are considered for time analyses. In addition, ProjectCost is associated with Input, which can support the calculation of life cycle cost.

Another important connecting point between life cycle assessment and cost or time analyses is through the definition of function. In life cycle assessment, the selection of a function unit is critical and determines the process of an analysis. On the other hand, the unit of resources in the construction management domain can be different from that in life cycle assessment. For example, a typical function unit for insulation materials is R-value in life cycle assessment. However, the unit for cost analyses can be in square foot. Thus, matching units between FunctionUnit and the BasicUnit attribute of ifcCostValue is important to guarantee a correct analysis.

Temporal and spatial considerations are reflected in three concepts, LCA-Study, InventoryList and UnitCost. The LCAstudy has a timestamp showing the

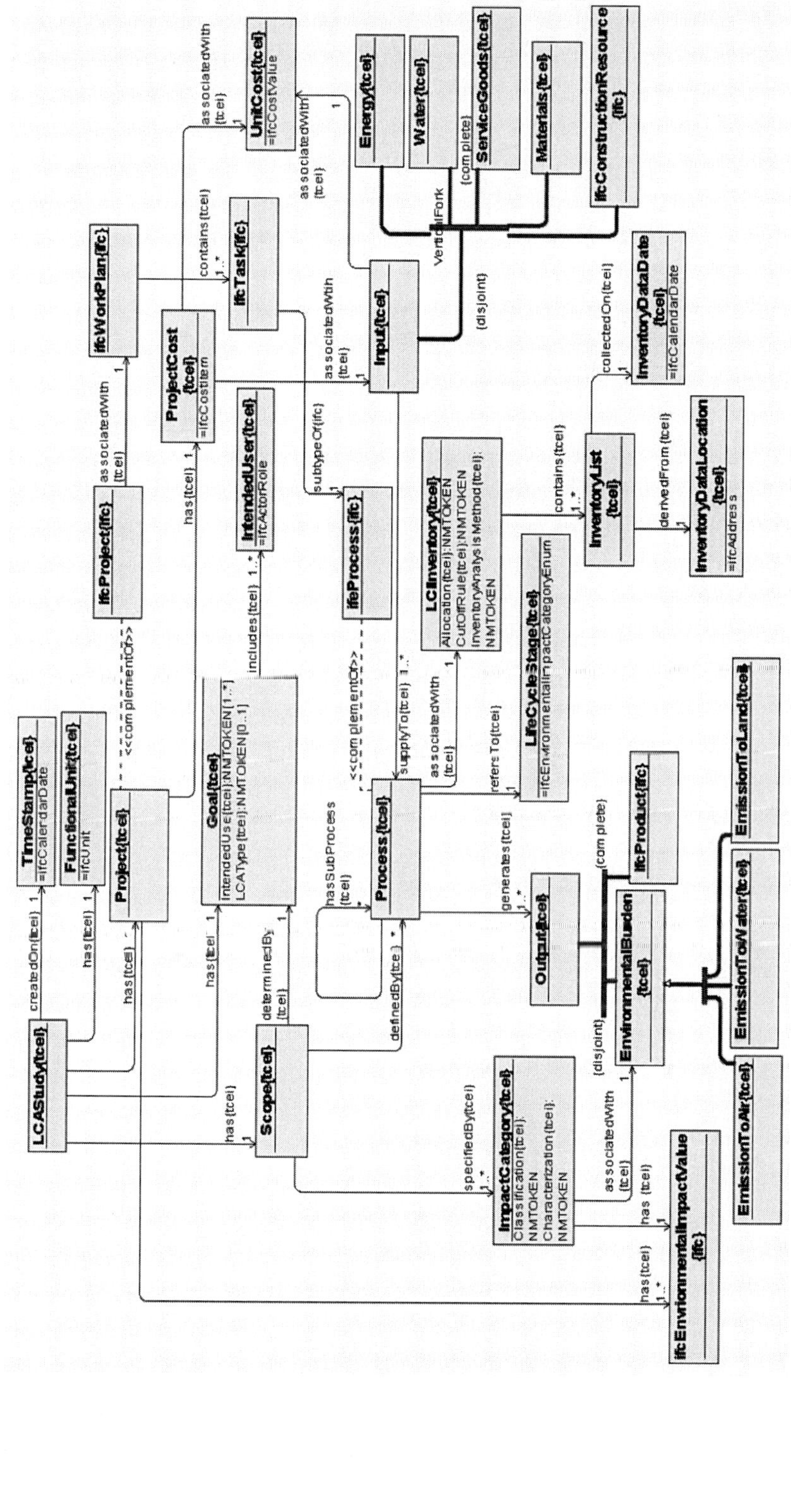

Figure 4-8. Sample ontology for integrated analysis

time when a study is conducted. Data used for environmental impact assessment or cost analyses should be reasonably close to the time when the study is conducted. Similarly, data should be representative and specific with respect to the geographic location that a study refers to. The temporal and spatial considerations in LCAStudy are reflected by TimeStamp and the address associated with ifcProject. On the other hand, both InventoryList and UnitCost are specifically associated with DataDate and DataLcoation. The temporal and spatial data help to determine the existence of mismatches among data. The information can guide designers or end users if adjustments to time, cost or environmental impact data are necessary.

Although the concepts are still at a high level, they form a framework for further development and can already support basic integrated analyses. There are only a limited number of concepts in the ontology that represent the construction management domain. This is because this domain has been well studied for decades and many data models such as IFCs can be reused. Therefore, only a few concepts are modeled for the sake of demonstration.

CASE STUDY

SimulEICon is a tool developed to perform time, cost and environmental impact analyses and select optimal solutions. Data used by the analytic procedure come from three sources, i.e., 1) building information models (BIMs) for geometry, quantities, and original material types of a design, 2) a pre-defined external database of additional material types and associated data such as costs, installation information and thermal performance, and 3) environmental impact data of building materials and construction processes. The use of the TCEI-Ontology can potentially identify or resolve discrepancies among time, cost and environmental impact data of different design options.

Figure 4-9 shows a sample case, in which generic component definitions are used for different elements in the building (such as exterior walls, interior partitions) in a BIM. On the other hand, in order to evaluate different options, other options such as wall types need to be incorporated for considerations. For example, Figure 4-10 shows other exterior wall types found in external sources. How can one decide what options from the list (Figure 4-10) should be considered as a replacement to the design choice (Figure 4-9) with comparability?

Table 4-5 shows three examples of exterior walls with materials and their properties. The thickness of expanded polystyrene in the steel stud wall design or the wood stud wall design, labeled as TBD in the table, may not produce the same R value as the SIP design. In order to make them functionally comparable, i.e., having the same R value, the thickness of the insulation in steel stud walls or wood stud walls may need to be adjusted. The R value affects not only embodied energy of insulation materials but also energy consumptions during

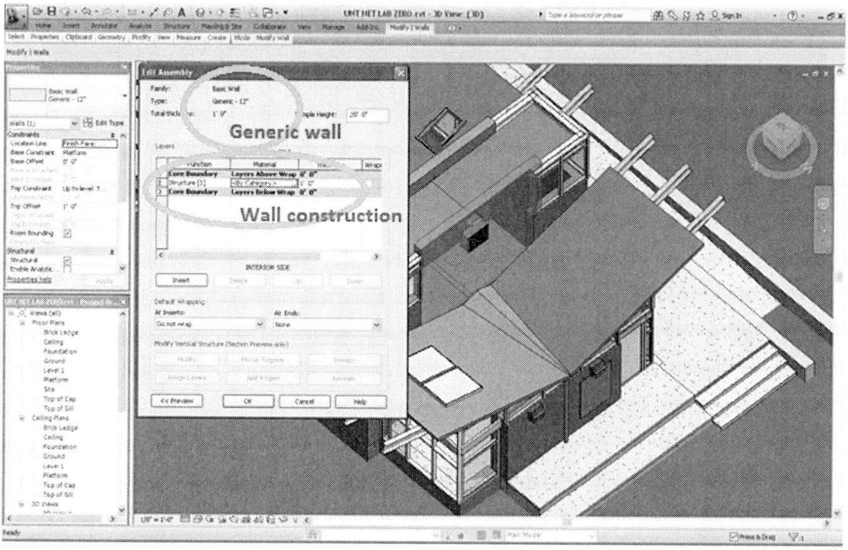

Figure 4-9. Sample case

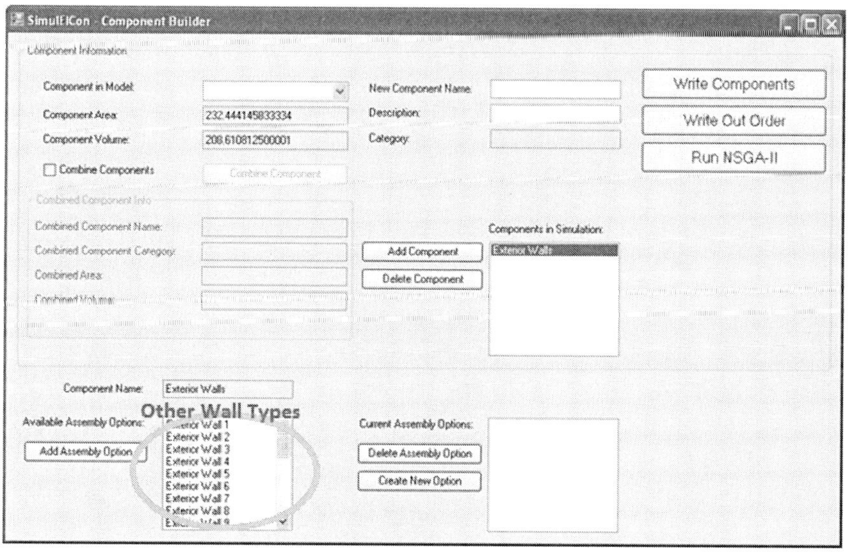

Figure 4-10. Options of other wall types

the occupancy stage of a building. Using TECI-ontology, FunctionalUnit in LCAStudy can track the functional unit used in each design. In addition, the scope of cost data can be represented by ifcProcess that cost data cover. Furthermore, the scope of greenhouse emissions can be represented by Process

Table 4-5. Examples of exterior walls

Component Alternative	*Material*	*Material's Properties*			*Total R-Value*
		Thermal Conductivity k (Btu(IT)/ h-ft-F)	*Thickness (inches)*	*R-Value*	
SIP 5.5″ thickness	Expanded Polystyrene	0.0225	5.5	20.3398	22
	OSB (Both face 7/16″ each side)	0.5258	0.875	1.0516	
Steel Stud Wall 1 5/8″ x 3 5/8″ Steel Stud at R-0.0031807/in (ASHRAE 2010)	1/2″ FR Drywall	0.1456	0.50	0.2862	TBD = 22
	3/8″ Plywood	0.252	0.38	0.3863	
	Expanded Polystyrene	0.0225	TBD	21.3275	
Wood Stud Wall Wood Stud 2x4 at R-1.25/in (ASHRAE 2010)	1/2″ FR Drywall	0.1456	0.50	0.2862	TBD = 22
	3/8″ Plywood	0.252	0.38	0.3863	
	Expanded Polystyrene	0.0225	TBD	21.3275	

or the construction stage of Process. Thus, when comparing the three options, TECI-ontology can be very helpful, especially when there are a lot of options.

SimulEICon can perform integrated data analyses. Genetic algorithms are used by SimulEICOn to select optimal solutions (Orabi et al. 2012, Zhu and Orabi 2012, Ozcan et al. 2011). Through a series of selections using extracted data from a BIM, in conjunction with available options stored in a database, results generated by the genetic algorithms are interpreted by SimulEICon and displayed in several formats. Details of each selected component option can be viewed as well as information on the entire project, including the estimated total project duration, cost, and environmental impacts.

For example, building assemblies of different designs can be compared side-by-side (Figure 4-11), including unit costs, unit carbon emissions and the composition of assemblies. Furthermore the unit cost and emission details of each material can also be tracked. To achieve this, it requires an integration of cost and environmental impacts. In addition, Figure 4-12 shows that the integration of time, cost and environmental impact allows designers to see a list of optimal solutions. For each design option, designers can track design components or compare different optimal solutions and their associated

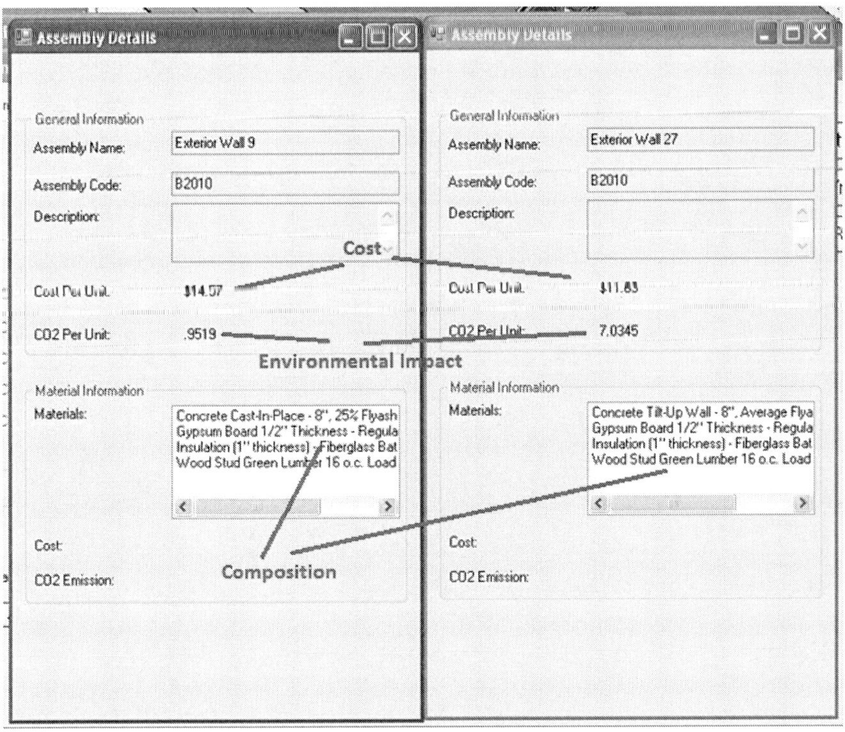

Figure 4-11. Example of comparing building assemblies

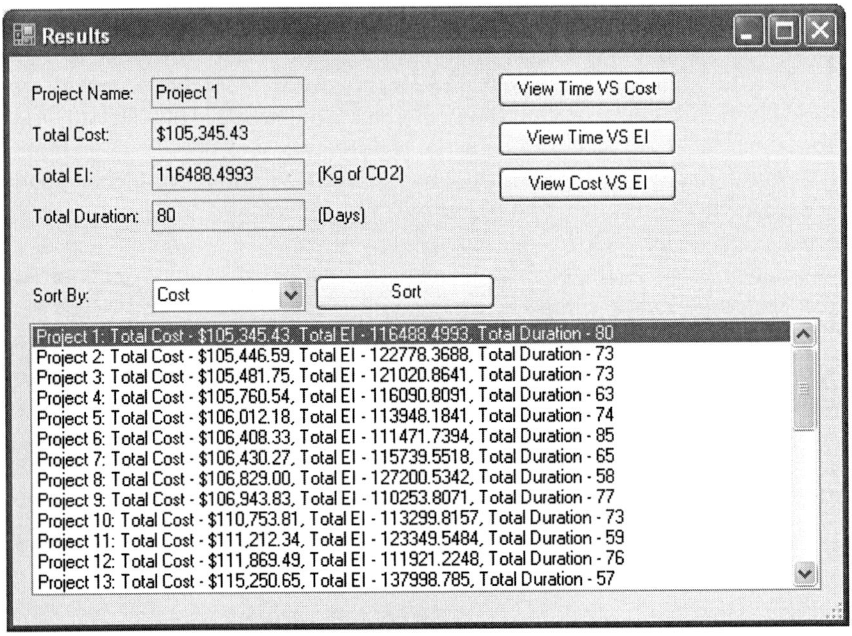

Figure 4-12. Results of optimal solutions

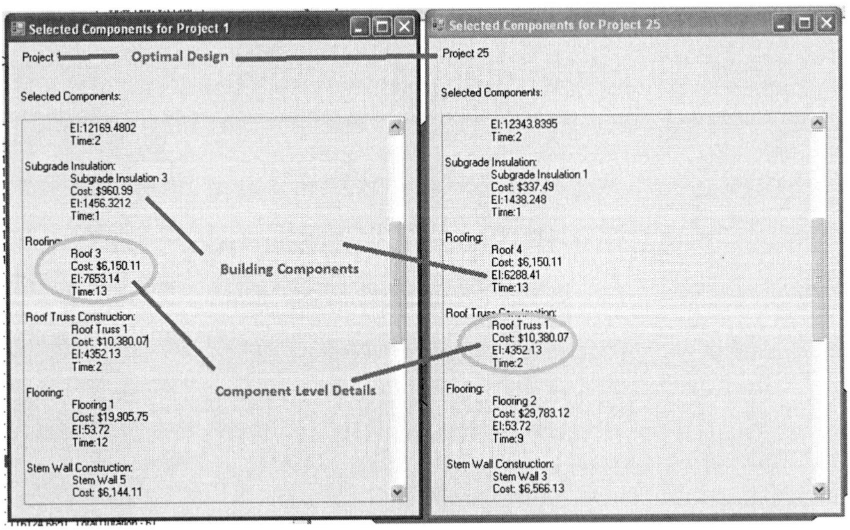

Figure 4-13. Details of optimal solutions

building components (Figure 4-13). Once implemented, the TECI-ontology can provide more functionality to support the design process, including for example providing the rationale of design decisions on the selection of each building element.

CONCLUSIONS

Construction projects are multi-objective. Many decisions to satisfy the objectives are made during the design phase of construction projects. In addition to traditional objectives such as project duration, cost, quality and safety, reducing environmental impacts is increasingly catching people's attention in the context of global warming and climate change, especially in research communities. However, data fragmentation is still a classic and open problem in the construction industry, even though there have been decades of research on this subject. The inclusion of environmental impacts adds another dimension of complexity to the subject.

To support integrated multi-objective analyses of design options, it is critical to understand and define the intent of each design option in order to track the compatibility of design options. It is found that key concepts such as goals, functions and scopes of a design are center pieces to define the intent of a design and backbones to form a coherent conceptual framework. Such a framework is essential, as shown in this chapter, to bringing together concepts of different categories, which define social, economic and environmental requirements, project performance criteria and date sources, and functions of a design. These concepts in turn provide a foundation to support the integration of data from different sources, better comparisons of design options, and searches for optimal design solutions. In those applications, ontology can be an ideal choice to represent the conceptual framework.

While ontology development on life cycle assessment is already ongoing, the integration of different types of data and knowledge for integrated analyses is not adequately addressed by the research community. In addition, as of this writing, there is no life cycle assessment ontology that is public available. This chapter shows that the idea of applying TCEI-Ontology to support integrated analyses of time, cost and environmental impacts has great potential. Especially, since issues such as data sources, data scope and functional units associated with life cycle assessment are so common, integration of LCA data with time and cost data just makes the issues more complex if there are no additional measures to alleviate the complexity. Thus, ontology as TCEI Ontology is important, not only at technical level to support effective data integration, but also at application level to allow professionals to compare different design options from multiple perspectives.

Although currently the scope and details of the proposed TCEI-Ontology are limited, it is extensible based on future application needs. Further development and tests are needed to determine the feasibility and the strengths and weaknesses of the proposed ontology. In addition, the purpose of this ontology is not to replace current ontology development efforts in construction or life cycle assessment. Rather, it is for filling gaps between different concepts such as cost and environmental impacts.

References

Anumba, C. J., Issa, R., Pan, J., and Mutis, I. (2008) "Ontology-based information and knowledge management in construction", *Construction Innovation: Information, Process, Management*, 8(3), 218–239.

Beetz, J., Van Leeuwen, J., and De Vries, B. (2009). "IfcOWL: A case of transforming EXPRESS schemas into ontologies." *Artificial Intelligence for Engineering Design, Analysis and Manufacturing*, 23(1), 89–101.

Bilec, M., Ries, R., Mattews, S., and Sharrard, A. L. (2006). "Examples of a hybrid life-cycle assessment of construction processes." *Journal of Infrastructure Systems*, 12(4), 207–215.

Bräscher, M., Monteiro, F., and Silva, A. (2007). "Life cycle assessment ontology." In *La interdisciplinariedad y la transdisciplinariedad en la organización del conocimiento científico: Interdisciplinarity and transdisciplinarity in the organization of scientific knowledge: Actas del VIII Congreso ISKO-España, León, 18, 19 y 20 de Abril de 2007* (pp. 169–178). Servicio de Publicaciones.

Froese, T. and Yu, K. (1999). "Durability of Building Materials and Components." M. A. Lacasse and D. J. Vanier (editors), Institute for Research in Construction, Ottawa ON, K1A 0R6, Canada, 2825–2835.

Gruber, T. R. (1993). "A transition approach to portable ontologies." *Knowledge Acquisition*, 5(2), 199–220.

Khasreen, M. M., Banfill, P. F. G., and Menzies, G. F. (2009). "Life cycle assessment and the environmental impact of buildings." *Sustainability*, 1, 674–701.

LEED (2010). *LEED pilot credit library: pilot credit 1: life cycle assessment of building assemblies and materials*, US Green Building Council, Washington, D.C.

Lima, C., El-Diraby, T., and Stephens, J. (2005). "Ontology-based optimization of knowledge management in e-construction." *ITCON*, 10, 305–327.

Lippiatt, B. C. and Boyles, A. S. (2001). "Using BEES to select cost-effective green products." *The International Journal of Life Cycle Assessment*, 6(2), 76–80.

Mao, W., Zhu, Y., and Ahmad, I. (2007). "Applying metadata models to unstructured content of construction documents: A view-based approach." *Automation in construction*, 16(2), 242–252.

Marzouk, M., Madany, M., Abou-Zied, and El-Said, M. (2008). "Handling construction pollutions using multi-objective optimization." *Construction Management and Economics*, 26(10), 1113–1125.

Matthews, H. S. and Small, M. J. (2001). "Extending the boundaries of life-cycle assessment through environmental economic input-out models." *Journal of Industry Ecology*, 4, 7–10.

Molenaar, K., Gransberg, D., Korkmz, S., and Horman, M. (2009). *Sustainable, high performance projects and project delivery methods: A state-of-practice report*, Charles Pankow Foundation, Vancouver, WA.

Moreno, A., Cappellaro, F., Masoni, P., and Amato, A. (2011). "Application of product data technology standards to LCA data." *Journal of Industrial Ecology*, 15(4), 483–495.

Noy, N. F. and McGuinness, D. L. (2001). "Ontology development 101: A guide to creating your first ontology." *Stanford Knowledge Systems Laboratory Technical Report KSL-01-05 and Stanford Medical Informatics Technical Report SMI – 2001 -0880*, (available at http://www.ksl.stanford.edu/people/dlm/papers/ontology-tutorial-noy-mcguinness-abstract.html).

Orabi, W., Zhu, Y., and Ozcan, G. (2012). "Application of multi-objective genetic algorithms to emission analysis of construction operations." *Construction Research Congress*, West Lafayette, May 21–23, 2012.

Ostad-Ahmad-Ghorabi, H., Rahmani, T., and Gerhard, D. (2012). "An ontological approach for the integration of life cycle assessment into product data management systems." *CIRP Design 2012*, 249–256.

Ozcan-Deniz, G., Zhu, Y., and Ceron, V. (2011). "Time, cost and environmental impact analysis on construction operation optimization using genetic algorithms," *Journal of Management in Engineering*, ASCE, 28(3), 265–272.

Pandit, A. and Zhu, Y. (2007). "An ontology-based approach to support decision-making for the design of ETO (Engineer-To-Order) products." *Automation in Construction*, 16, 759–770.

Rezgui, Y. (2006). "Ontology-centered knowledge management using information retrieval techniques." *Journal of Computing in Civil Engineering*, 20(4), 261–270.

Rezgui, Y., Boddy, S., Wetherill, M., and Copper, G. (2011). "Past, present and future of information and knowledge sharing in the construction industry: Towards semantic service-based e-construction?" *Computer-Aided Design*, 43(5), 502–515.

Schevers, H. and Drogemuller, R. (2005). "Converting the industry foundation classes to the web ontology language." *Proceedings of First International Conference on Semantics, Knowledge and Grid, 2005, (SKG'05)*, 73–73. IEEE, 2005.

Sharrard, A. L. (2007). *Green Construction Processes Using an Input-Output-Based Hybrid Life Cycle Assessment model.* Ph.D. Dissertation, Carnegie Mellon University, Pittsburgh, PA.

Singh, A., Berghorn, G., Joshi, S., and Syal, M. (2011). "Review of life-cycle assessment applications in building construction." *Journal of Architectural Engineering*, 17(1), 15–23.

Staub-French, S., Fischer, M., Kunz, J., and Ishi, K. (2003a). "A feature ontology to support construction cost estimation." *Artificial Intelligence for Engineering Design, Analysis and Manufacturing*, 17, 133–154.

Staub-French, S., Fischer, M., Kunz, J., and Paulson, B. (2003b). "An ontology for relating features with activities to calculate cost." *Journal of Computing in Civil Engineering*, 17(4), 243–254.

Suh, S., Lenzen, M., Treloar, G. J., Hondo, H., Horvath, A., Huppes, G., Jolliet, O., Klann, U., Krewitt, W., Moriguchi, Y., Munksgaad, J., and Norris, G. (2004). "System boundary selection in life-cycle inventories using hybrid approaches." *Environmental Science Technology*, 38, 657–664.

UNEP (2011). *Towards a life cycle sustainability assessment: Making informed choices on products*, United Nations Environment Programme, Nairobi, Kenya.

Yudelson, J. (2009). *Green building through integrated design*, McGraw Hill, New York.

Zhang, Z., Wu, X., Yang, X., and Zhu, Y. (2006). "BEPAS – a life cycle building environmental performance assessment model." *Building and Environment*, 41, 669–675.

Zhong, B. T., Ding, L. Y., Luo, H. B., Zhou, Y., Hu, Y. Z., and Hu, H. M. (2012). "Ontology-based semantic modeling of regulation constraint for automated construction quality compliance checking." *Automation in Construction*, 28, 58–70.

Zhu, Y., Pandit, A., and Mao, W. (2006). "Framework to support the representations of semantic mapping for a hybrid integration strategy." *Journal of Computing in Civil Engineering*, 20(4), 237–246.

Zhu, Y. and Orabi, W. (2012). "An optimization model of time, cost and carbon emissions for building design and planning." *Proceedings of the 1st International Conference on Building Sustainability Assessment*, Porto, Portugal, May 23–25, 2012.

CHAPTER 5

From Deep Blue to Watson: The Nature and Role of Semantic Systems in Civil Informatics

T. E. El-Diraby*

Abstract: *There is no such thing as an isolated piece of knowledge. Networks of unstructured information sources are created and linked online daily. The proliferation of such extended sources of knowledge has caused a shift in the design of informatics systems from rule-based approaches (which dominated formal ontology) to statistical or mining-based approaches. Such shift is clearly manifested when we compare the nature and structure of IBM's Deep Blue computer (which relied on procedural and logical steps to reason about Chess) and the new Watson computer (which relies on probabilistic and semantic analysis of unstructured web-based material).*

It is argued that changes to ontology development are needed to meet the challenges and make use of the opportunities of extended online sources of knowledge. This starts by shifting the epistemological foundations of modern informatics ontology from positivism to constructivism, and replacing foundationalism with coherentism. The traditional desire for using ontology in reasoning should be complemented by an equal belief in the value of the representational role of ontology. Equally important, the belief in the role of ontology as a (static) formal encapsulation of knowledge should be balanced with a new belief in the role of ontology in discovering and integrating new knowledge from extended sources. Axioms must be limited and geared towards listing what is rejected instead of enumerating what is true. Entity structures must be flatter and closer to a network format instead of the tree-like format of taxonomies. The ontology must also interact with a set of auxiliary services that are needed to help customize ontology elements to match the context of its application, integrate new discovered knowledge; and for self-healing of axioms with new acquired knowledge.

*Associate Professor and Director, Centre for Civil Informatics, Dept. of Civil Engineering, University of Toronto. 35 St. George St., Toronto M5S4A1 ON Canada. tamer@ecf.utoronto.ca.

In proposing these, and in true adherence to coherentism and constructivism, the author is not advocating that the implementation of these tools is the correct approach or an approach that is applicable to all situation.

INTRODUCTION: WINDS OF CHANGE

e-society initiatives (of the 1990s) aimed to put people "on line" to surf the web. Today, evolving k-society (knowledge society) initiatives promise to put people "in power". Thanks to social media (Web 2.0), the web is morphing into a new socioeconomic space where e-citizen and e-democracy concepts are thriving. Social networks are forming constantly to promote causes and coordinate events (as small as a party and as big as the Egyptian revolution). Beyond sharing media, the essence of k-society is to democratize innovation and empower people harness "collective intelligence" in order for them to lead knowledge-enabled decision making (von Hipple 2005; Smart 2010b).

Unstructured sources of knowledge are being added to the web constantly by professional bodies and general public. Web users post their knowledge formally in the form of articles, manuals, datasets, and wikis, to list a few. Indirectly, some of their knowledge is embedded in their social web signature: their tags, comments, discussions, and tweets. As chaotic as theses corpus of data may seem, we can use advanced tools to harness some of the knowledge in them. In fact, crowdsourcing and analysis of *ad hoc* online comments is evolving as an enabler for reverse innovation. Even more, analyzing the very structure of the social network of a user can help reveal some of the implicit knowledge related to the domain of this user. In the simplest form, by studying the linkage between different users and comparing these to their semantic profiles, we may deduct ontological relationships between the domains of knowledge of these users. By studying the patterns of communication between them, we can also study synergies between their domains.

Beyond collecting and analyzing consumer comments, harnessing the knowledge of the crowd is reshaping some of the professional practices. As a case in point, Goldcorp shared 400 gigabytes of its data with the crowd and announced a prize of $575k to anyone who could help discover more gold. Receiving 1000 entries, they discovered gold worth $6 billion. Crowdsourcing is also fostering the creation of new business models. For example, *InnoCentive*, a problem-solving marketplace has 250,000 "solvers" competing for more than $35 million in prizes.

The *ad hoc* and unstructured nature of big, dispersed and heterogeneous data corpus contrasts the formal and static nature of ontology-based systems. The introduction of ontology to informatics systems (at least in Engineering) was mainly done by researchers with backgrounds in artificial intelligence (AI). The aim was to formalize domain conceptualization (Gruber 1995). i.e. in contrast to simplified and contextualized claims of a rule-based expert systems, AI ontologists were trying to establish axioms about abstract concepts that are supposedly universally applicable. Tools such as first order logic or predicate calculus were used to make sure ontologies are consistent and are machine-interpretable.

We have to explore *neo* roles of ontology to match the evolving trends of k-society. The current (AI-inspired) mentality aimed to support a single decision maker to "reason" about "technical" issues. This mentality preferred the use of top-down approaches to build formal algorithms in relatively very static structure. Rather, today's informatics systems have to deal with a work environment that includes distributed decision makers, who are struggling to "integrate" technical and non-technical aspects of project with a milieu of unstructured sources of knowledge. In the era of big data, such knowledge is encapsulated (directly or indirectly, intentionally or not) in data and information corpus that are spread in a complex web. Such knowledge is multifaceted, where many views of the same concept can co-exist. Knowledge is also dynamic, where the representation of the same concept can change from one context to the other.

The need to re-engineer ontology role, its structure and development means is also warranted to match the evolving nature of practice in civil engineering systems. These systems are no longer just physical artifacts. They are now an integral part of their social fabric. Decision makers (public officials, planners and engineers) need to balance traditional technical analysis, public safety, service delivery, and maintenance costs. They must preserve biodiversity, conserve energy and assure affordability. Recently, they also have to assure that the system ties to its socio-economic environment and addresses issues such as job creation, aesthetics, urban form, and local lifestyle. Consequently, ontology cannot just continue to be a first order or predicate calculus artifact. It has to incorporate more linguistic and text analysis systems to accommodate the subjective nature of the topics above.

This chapter is, first, an attempt to observe some challenges and opportunities for the work on research in ontology (and informatics in general), re-investigate the role and purpose of ontology in civil informatics, and suggest approaches (new tools) that can help enhance the relevance and contribution of ontology in the larger context of civil informatics. Adopting a constructivist epistemology, the analysis below and the proposed way ahead/solutions are but a "fallible" proposal that is meant to showcase one possible approach and is aimed more to start a debate rather than charter a certain way forward.

WATSON: THE ERA OF EXTENDED MIND & UNSTRUCTURED DATA

Emerging in philosophy and because its name refers to modeling what exists (or being), there is a rather false expectation that ontology should be exhaustive in scope and inclusive in its terminology. Nothing can be further from the truth. Ontology is a just a temporary and relativistic claim for knowledge. In fact, constructivists deny the existence of universal ontology altogether. "Knowledge is a shallow stream, its depth nowhere is the same; communication is always an unaccomplished task, and philosophy lives in argument, not in agreement. Scepticism cannot be expunged, for it is philosophy's source of question . . . the very hope that the ontological options could be reduced to just one, even for the purpose of a given discourse is antiphilosophical—not to say sophistic (Swindler 1991)".

In contrast to the structured, top-down, and holistic view of rule-based philosophy, the extended mind thesis (EMT) adopts an *ad hoc*, distributed and unstructured view of knowledge, which could be more suitable to today's socio-technical culture. EMT makes two fundamental claims. First: cognitive processes (ones that take place within an agent's mind) such as retrieval of memories and linguistic processes can be partially constituted by portions of the world that are not bound by their brain-and-body. Second: an agent's non-occurrent beliefs can be constituted (at least partially) through sources that are not bound by their system. An agent here refers to a knower system, which could be a human being (with sets of beliefs) or an ontology-based system with a set of axioms. Four conditions must be met to accept non-occurrent beliefs form extended sources (Vold 2011):

1. Constancy: The use of the resource must be a constant in the agent's life.
2. Accessibility: The resource must be easily analyzed.
3. Reliability: The agent must trust the resource.
4. Endorsement: the information in the resource must be there as a consequent of having been consciously endorsed at some point.

To illustrate, if we ask an agent or a knower system: "what do you think about design-build project delivery system?" Assuming that its existing rules (beliefs in the case a human knower or axioms in the case of an ontology-based system) state that "it is a project delivery system typically used in fast track projects", the agent will respond accordingly. If we further ask "how about design-build in non-fast track projects" and, assuming that, the rules of the knower system are silent about this question, then it will be helpful if the agent is presented with evidence that the local department of transportation (DOT) has approved design-build in 40% of its non-fast track projects compared to 70% usage in fast track projects. The endorsement of the DOT as a knowledge authority should allow the knower system to answer: "design-build is usable in non-fast track projects". The challenge increases if we then ask the agent about integrated project delivery (IPD), which is not defined in the rules of the knower systems. We will not get an answer unless this knower system is linked to an extended pool of knowledge such as DOT web site or its procurement manual or the American Institute of Architects, whereby it can find that the DOT defines it as "is a collaborative alliance of people, systems, business structures and practices into a process that harnesses the talents and insights of all participants to optimize project results, increase value to the owner, reduce waste, and maximize efficiency through all phases of design, fabrication, and construction (AIA 2013)". Such piece of knowledge does not violate any of the ontology axioms and comes from a reliable extended source—it is therefore contingently affirmed.

The rule-based thinking, which was used in simple formats in expert systems or in philosophical/abstract format in the form of ontology axiom, is being challenged by what can be called mining-based or statistic-based system. That is, our need is no longer that the computer applies a set of static/formal rules to help us reason about a complicated problem. Rather, the need is that the computer system helps us distil knowledge widely spread in an unstructured format all over the web (Qvartrup 2006). For example Chunara at al. (2013) compared posts on Facebook and local health data to elements and patterns of obesity in several regions!

The shift from rule-based to statistical-based systems is evident from IBM mega challenge work agenda. In the 1990s, IBM invested heavily in developing Deep Blue—a computer system that used artificial intelligence with sophisticated mathematical/logical rules to master Chess and defeat the world champion (Ksparov). In contrast, recently, IBM massively invested in Watson—a web-based system to search internet corpus and use statistical analysis and approximate reasoning means to find possible knowledge. It competed against humans in Jeopardy (the popular American TV show) and won.

Our challenge is to use advances in k-society and the proliferation of extended sources of data to develop informatics systems that can facilitate the integration of such unstructured sources with the stable and structured nature of ontology. Can we reach these sources, assess their relevance, examine their validity, and test their coherence to our ontologies? If proven valid, can we distil knowledge from these data sources and embed it into our informatics system in a manner that observes the contextual requirements of knowledge management? (for more, see El-Diraby 2011).

To do so, we need to carefully consider the morphology of data, information and knowledge, re-examine our views and definitions of the role of informatics systems in today's environment, and re-evaluate the components, nature and methods of ontology development. The following sections are an attempt to suggest a roadmap to do just that.

First, we may want to examine the nature and role of ontology. Ontology is a branch of philosophy that focuses on the science of what is, of the kinds and structures of objects, properties, events, processes, and relations in every area of reality. Sometimes "ontology" is used in a broader sense, to refer to the study of what might exist. In informatics, however, the main use of ontology (at least initially) was related to classification (analysis of entity types) and on constraints on allowable taxonomies (Smith 2003). The aim is to unify the semantics and terms used (in databases or product models). Gruber (1995) shifted the focus to domain "conceptualization". What are the main concepts (universal) concepts in a domain: "A conceptualization is an abstract, simplified view of the world that we wish to represent for some purpose. Every knowledge base, knowledge-based system, or knowledge-level agent is committed to some conceptualization, explicitly or implicitly (Gruber 1995)". This supported a reconsideration of the tools and means for developing ontologies including a wide (but rather selective) range of ontological models, theories and frameworks for such conceptualization. CYC, for example, utilize the typical philosophical notions such as identity, part-whole, set membership (Guarino and Welty 2000).

THE CHALLENGE: FROM DATA TO INFORMATION TO KNOWLEDGE

Data is a representation of the simplest facts about a system with limited meaningfulness. In information systems, data is normally stored in databases. Information is the composition of various data to establish a meaningful representation of facts. In information systems, information is exchanged normally through communication between humans or via electronic means such as web sites or e-mail (for more, see Floridi 2004). Typically, IT-based tools (such as XML and other web systems) are used to support the interoperable exchange of information—for example, IFC (industry foundation classes).

Knowledge is the wisdom gained from experience and understanding of the inner behaviour of systems that enables better decision making. While data and information tend to be technology-oriented, knowledge is basically human-centered. Knowledge is "a fluid mix of framed experience, values, contextual information, and expert insight that provides a framework for evaluating and incorporating new experiences and information. It originates and is applied in the minds of the knowers (Davenport and Prusak 1998)". Lately, knowledge definition progressed to include more of its societal nature (Stehr 2001). Knowledge is a social phenomena not merely a psychological phenomenon (an entity held in the mind of an individual). Cognition emerges with environment, where learning/understanding is embedded not only in the minds of individuals or in knower systems, but, equally important, in work processes, and the conduct and relationships among humans (Smart et al. 2010).

k-society embraces the extended mind thesis (EMT), where outside objects and knowledge sources act as a "coupled system" (March and Onof 2007) or an interactive 'ecology' of knowers (App-empowered people) and their social norms and networks (Davenport and Prusak 1997). An open (constructivist) model of innovation prevails in this ecology—innovation is created within and outside the organization. All participants can dynamically assume the role of creators, users or brokers of knowledge (Grant 1996). Such knowledge has three main features:

Knowledge is Distributed

Analysis of civil engineering systems is multidisciplinary. It is impossible to find all required expertise in one place. A diversified knowledge supply chain has to be weaved to address such analysis (Hastingsy 2005). Such knowledge is, essentially, distributed.

Knowledge is Contextual

Knowledge representation only makes sense when located in "contexts". A "context" is one of the possible worlds generated by adopting a situational perspective where "concepts" portray different behaviour, triangulate in different semantic networks and obey modified axioms (McDermott 2001).

Knowledge is Social

KM tools have to support/promote the social nature of knowledge as it emerges through human interaction (whether face-to-face or virtually through social media).

Due to the above changes in culture and work environments, research in informatics systems faces new challenges that requires changes in the way ontology design and management. First, informatics relates to tools of work of an era where information (not data) is the starting point of action. In other words, the use of advanced ICT tools has made access to and manipulation of data (including real time data) so prevalent that it is a given. Unlike the 1990's (when informatics ontologies research started), the aim is not to just collate scattered data into information. Rather, it is to use easily available information to formalize/represent knowledge. So, while initially it was trying to use *data* to support *information* development and usage, informatics is now trying to use the wealth of available *information* to support *knowledge* management.

Further, in the 1990's there was a considerable separation between producers and consumers of information management systems. The consumer was, mainly, a passive one. The objective of researchers was an attempt to bridge the gap between the established business management practices and the, then, new field of information management systems. In other words, companies had well established work processes. They wanted to complement that with new information services that were seen as an add-on to enhance efficiency and tracking of performance. Now, information is at the core of enterprise assets. In the knowledge economy, virtual enterprises emerged to provision systems and deliver solutions solely through web-based communications. "Knowledge products" (software and web services) are flourishing as the most valuable commodities in this economy. The transformation of IBM from a hardware (data crunching) manufacturer into a business solution provider (with knowledge management as core competencies) is a case in point[1]. In fact, informatics "achieves" the organization—leading web based organizations such as Google, Facebook, and Amazon.com are essentially processors of big data, managers of information flows, and leaders in distilling knowledge and business intelligence from unstructured web corpuses.

Informatics does not serve the enterprise only. It aims to reach out to and exploit the knowledge of the whole society. It does not deal with a divided producers and consumers of knowledge—they have both morphed into "prosumers", a portmanteau formed by contracting the word professional (or also producer) with the word consumer. It is used to signify the active role of consumers in producing the products they use.

---

[1]Another example is the demise of Sun Microsystems and its acquisition by Oracle

In short, in the early stages of informatics research, computation and management process were in parallel. The objective was to bridge them. In the current time, there is no gap between information and management processes. Information is the main asset not only for the enterprises but also for the whole society. Hence, the prevalence of ideas such as open source, the network as the computer, service-oriented architecture, cloud computing, and enterprise 2.0.

Finally, the initial scope of informatics work includes a composition of computational/automation systems (of databases, AI, knowledge bases, etc.) that aimed to process data in order to support formal decision making. Informatics is oriented towards human empowerment: allowing people to create and share data (open access & distributed data), integrating unstructured data to support the meaningful generation of new data or knowledge (remixabaility), and providing platforms for collaborative development and re-use of new/existing resources to innovate new "knowledge products". Therefore, informatics is more interested in linguistic, communication, and social issues—hence the introduction of semantic and social webs and the interest in the philosophy of information.

Summary

The evolution of social and semantic web technologies and their ever-increasing role in socioeconomic activities and human life have mandated a shift in the design of artificial knower systems. Traditional computational work started from data with the aim of making it meaningful *information for the enterprise.* Informatics starts with information with the aim of making it meaningful to the *knowledge society.* Informatics systems aim to address knowledge management issue beyond the simple computational dimension and with acknowledgement of (focus on) the human and social aspects of knowledge. Mainly, integration of human-based (less mathematical) means of knowledge processing such as linguistics and communicative aspects, (personal) utility/experiences and preferences, contextual awareness into these systems.

A POSSIBLE WAY FORWARD

The advancement of k-society is changing the landscape of knowledge management (KM). Innovations and new knowledge are no longer limited to organizational boundaries. Instead an open model where innovation takes place throughout the society prevails nowadays. Open innovation democratizes information by enabling users, partners, governments and other stakeholders to participate in the design, development, implementation and management of new products, services, and solutions (von Hippel, 2005).

In light of this, the following sections present a set of changes that should shape the development of ontology and informatics systems. This covers the metaphorical role of ontology, the extended nature of ontology, the structure of axioms, the style of conceptualization, and the ontology development and

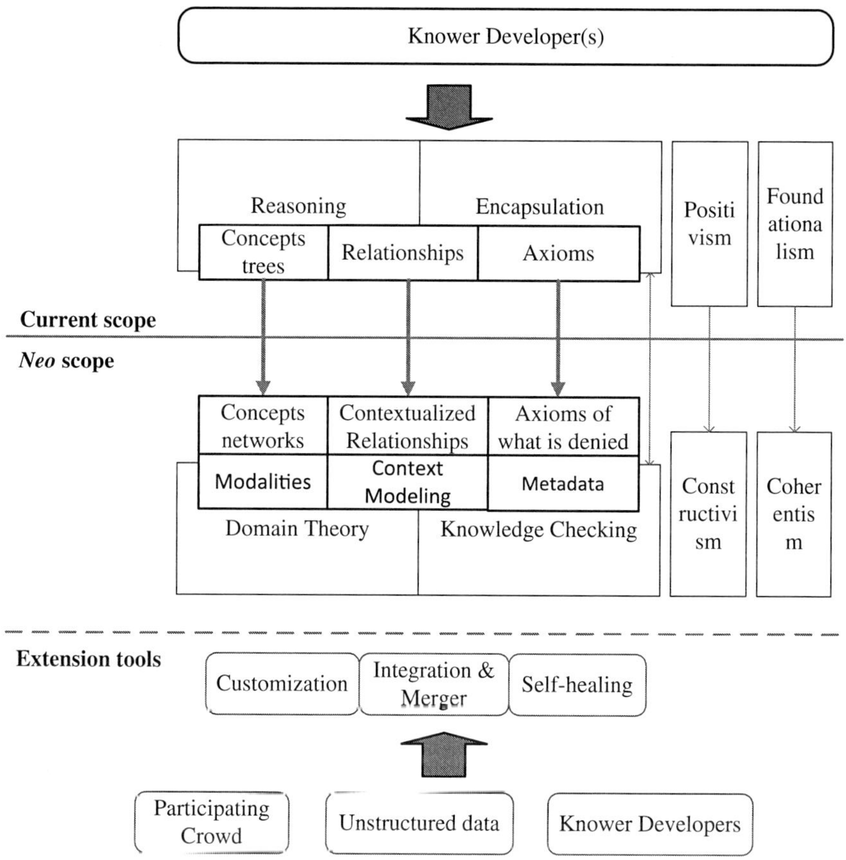

Figure 5-1. Ontology—before and after

validation methodology. Figure 5-1, is an attempt to show the current scope of ontology development which include concept trees, relationships and axioms. This scope is dominated by the desire for using ontology in reasoning and the belief that the only role for ontology is the (static) encapsulation of current knowledge. Figure 5-1 shows a neo structure that emphasis the development of concept networks (in contrast to concept trees), relativistic and context-aware relationships (in contrast to generic relationships), and axioms that focus mainly on listing what is denied. Along with reasoning, this new scope argues that ontology is also a domain theory. Along with encapsulation, this scope argues that ontology is also a tool for checking and integrating new knowledge. At the epistemological level, this will require a shift from positivism to constructivism and from foundationalism to coherentism. Finally, a set of auxiliary services (shown at the bottom of Figure 5-1) are needed to help customize ontology elements to match the context of its application, integrate new discovered knowledge, and self-healing of axioms with new acquired knowledge.

EPISTEMOLOGICAL ASSUMPTIONS

Influenced by the positivism of AI and the desire for consistency and regress to enable sound reasoning, current ontologies, essentially adopt foundationalism. In this view, there is firm belief in a unique and stable grounding to all our knowledge which can be decomposed into simple rules that, in turn, can be aggregated to create other rules/facts using (first order) logic. The result is a tree-like structure of facts. Concepts have static and clearly defined behaviour and this can neither be doubted nor changed (contextualized). Coherentism on the other hand is more appropriate to the constructivism of k-society and the networked nature of extended and unstructured data. In this view, beliefs are linked to each other in a web. Each belief complements and strongly coheres to nearby beliefs, but not necessarily to beliefs far away in the web. Such a loosely-coupled behaviour is more adaptive to fuzziness and dynamic evolution of knowledge (see Gruninger and Fox 1995; Dumonter and Hoehndrf 2010).

RECONSIDERING THE ROLE OF ONTOLOGY

At the most fundamental level, we have to reconsider the role of ontology in supporting advanced informatics systems. Why do we develop ontologies and what is the scope of their functions? The following discussions are relevant in this debate.

Ontology as Reasoning Tool vs. A Domain Theory

At the most basic level, we need to reduce the emphasis on reasoning and provide equal focus on representation of the essential elements of knowledge. In other words, upon developing an ontology, we have to dedicate equal attention to developing a computational models (or a reasoning engine) and the investigation and establishment of a domain theory. The difference between the two is significant. Typically, a computational model is an AI-influenced artifact that focuses on the formal codification of facts and their use in regress analysis through automated means (such as a reasoner). Researchers have to use first order logic or predicate calculus to assure a stable structure and consistent inference—for more information, see Borgo (2007). In contrast, domain theory (which is a branch of mathematics) tends to focus on specifying denotational semantics through formalizing the intuitive ideas of approximation and convergence (Abramsky and Jung 1994). Domain theories are related to concept topology and tend to emphasis the conceptualization and epistemological aspects of domain knowledge beyond computational issues.

However, overemphasizing the domain analysis can lead to excessive abstraction (some can argue, reductionism) that is suitable for a philosophical ontology in contrast to an informatics ontology. Yet, not considering the epistemological foundations of an informatics ontology can reduce it to a reasoner (see Vissar and Bench-Capon 1997).

Concept Modeling vs. Context Modeling

The famed linguist J. R. Firth asserts that "you shall know a word by the company it keeps" in reference to the contextual nature of meanings. Context modeling has been rigorously used in information systems (Spink and Park 2004; Stuckenschmdt and Wache 2000), process reengineering (McDermott 2001), and subject-oriented programming (Harrison and Ossher 1993). It is gaining momentum in social web/media (Santanche et al. 2007), bioinformatics (Strang and Linnhoff-Popien 2004), and emergency management (Sotoodeh 2009). The need for contextualized knowledge representation has been highlighted by researchers in context-sensitive design (Stamatiadis 2005), and infrastructure interdependency (Tofani et al. 2010; Rosato et al. 2008). However, context modeling in civil and construction research has focused on the area of virtual reality, where a context refers to the physical parameters.

Much of the research on ontology in civil engineering has been consumed in developing models of concepts and their behavior. Limited work, however, has been dedicated to context modeling. How does the behaviour of concepts morph based on the needs of certain contexts. Concept features vary based on the context where they exist. For example, in a design-bid-build context, an engineering firm normally assumes the role of a designer (making and certifying designs). In a design-build context, an engineering firm could assume the role of project manager. During project design or construction, engineering firms could be involved as decision makers, analysts, or advisor. As such, the assertion "engineering firm has the role of designers" does not hold in all contexts. "It is not the mere problem of identifying an entity in the world that is central to the ontological representation of the world, but the ability to re-identify an entity in all its possible forms, or more formally re-identification in all possible worlds (Bench-Capon 1990)."

Ontology as Encapsulator of Knowledge vs. A Checker of Knowledge

Traditional views hold that an ontology is formal, static encapsulation of knowledge. This may mandate that ontology exhaustedly list all elements of knowledge (concepts, relations and axioms). However, the amount of knowledge that we can embed in an ontology is limited compared to what can exist in extended sources. So, it is important that we add to any new ontology tools to link it to *ad hoc* or extended sources of knowledge. We should model and embed means to check for authenticity and for consistency of new/extended sources. If acceptable, they should be (late) binded into the ontology. This means that ontology structure should focus on finding adequate and flexible ontology models that captures the essential essence of knowledge in the AEC domain. For example, instead of attempting to list all types of doors and their attributes, we should focus on modeling what is essentially correct about doors: how they behave, how they link to other elements physically, their cost, schedule and construction characteristics. Equally important, is to study and explore the logical and usage patterns of doors. If we are successful in reaching a succinct and inclusive model of the concepts under various circumstances and based on different scenarios of needs, then we can accommodate future new forms of doors (physically) and logical features. Ultimately, ontology should balance the encapsulation of knowledge and the checking the suitability of new pieces of knowledge and supporting their integration into its structure—at least contingently.

Ontology as an Enumeration of What is True vs. A List of What We Deny

Extended sources may include new knowledge that was not available to us. Our mentality in modeling concepts and their axioms within an ontology should shift from listing what is true to a limited list of what must be denied. i.e. we should focus on coding what is absolutely unacceptable rather than enumerating all that is true. This way, if a new piece of knowledge is discovered, it will be assumed to be relevant or worthy of consideration as long as it does not violate what is denied in the ontology. For example, we should focus on listing and describing different project delivery systems with a mind that is open to new variations. i.e. we cannot assume that the current modes are they only "true" ones. To open the door for new forms of project delivery systems, we should just focus on modeling what cannot be true and then judge new ideas (for example, integrated project delivery system) based on violating these rules.

THE EXTENDED ONTOLOGY

We have to embed within ontology means to help extend its usage beyond its contents. This includes three main features:

Interaction with Metadata

To facilitate the use of extended and unstructured data, ontology should include a section to describe and model metadata of extended knowledge sources. Each domain ontology must include within it means to develop metadata for its own concepts as this will enhance the usability of these concepts by other systems. At the same time, ontology should be able to describe metadata of acceptable or relevant pieces of knowledge. That is, ontology should be able to describe itself and be able to expose that to other sources, hence facilitating the analysis of similarity and relevance between these sources and the ontology. Roussey et al. (2011) show examples for the use of RDF and Dublin Core metadata in this regard.

Metadata can be instrumental in establishing trust. This is very important in assessing the quality of unstructured data—especially sources provided on the web. By investigating the social network of the users of a data source, we can learn a lot about its quality. Using variations of the Google page rank algorithms, online search engines prioritise results based on the level of social activities associated with them (frequency of commenting, linkage and usage of proposed data or even the degree of closeness of the people who already have used such data. If a page *A* has pages *B1* through *Bn* pointing to it, *e [0-1]* is a damping factor, and if we also define $C(A)$ as outgoing links from A, then the relative importance (or trust factor) of A is measured by the probability $P(A)$: $P(A) = (1 - e) + e(P(B1)/C(B1) + \ldots + P(Bn)/C(Bn))$. To model $P(A)$, we can use Markov chains or Particle Swarm Optimization, which uses a socio-cognitive model of influence and learning (Kenndey et al. 2001). We can use two formats: simple models, where the network structure is neglected (Granovetter 1978) and game theoretic models of contagion in which each individual looks at the pay-off for embracing a behavior or not.

Micro-formats (such as hCard) are simple descriptions of data structures which can be used also to embed metadata into unstructured data corpuses, people, places and organisations. Similarly, XFN is a social network related micro-format (Bao et al. 2009). To assess and communicate trust, OAuth is a mechanism to authenticate websites allowing them to exchange data in a trustful manner and increase the confidence of the user community, which should reduce the friction associated with the integration of related online services. Further, advances in data obfuscation allow for protecting information while sharing it. It allows providers of unstructured data to reveal information and manage inferences in a form that is appropriate to the degree of trust in the recipient (Bao et al. 2009).

Customization

One of the most important tasks in informatics systems is to enable the creation of customized forms of knowledge already contained in the ontology to accommodate the needs of specific contexts. The overwhelming rates of creation and flow of unstructured data on the social web have helped forge many usable systems for filtering and customizing knowledge flows. For example, FriendFeed is a tool that can help aggregate unstructured data across multiple sites, enabling contextual responses to search queries to be made. In the same vein, Mozilla Raindrop focuses on fragmenting a stream of disparate information into separate contextual pieces (Braines 2010).

While classical first-order logic has a universal nature, non-classical logics such as modal and temporal logics provide a powerful basis for modeling contexts, especially in socially-oriented modeling of processual situations (Chellas 1980). Modal logic deals with the description of concept modalities—normally through probabilistic analysis of possibility (variety of shapes or modes of existence) and necessity, where the meaning of expressions depends on an implicit context. Temporal logic deals with the dynamic change in logic constructs across time. The theoretical challenge is to use a ontology to generate concept possibilities to match the representational needs of a context. i.e. if E is a concept in our ontology $\bar{O}$. Can we find ways (such as deontic logic) to develop different modalities of E (mE) that describe varying behavioural patterns? We have also to investigate conditions (*Cn*) under which a modality *En* (for example, known project delivery system, new project delivery system, acceptable project delivery system) is possible (denoted $\Diamond En$) or is necessarily the case (denoted $\Box En$). When *En* is realized contingently, the causes (*Un*) of this could be accidental or non-accidental. *U* can also be known or unknown. Known *U* can be controllable or non-controllable. Of course, *C*, *En*, and *U* all have likelihoods (*L*). At time *t*, *En* could be related to other ontological concepts. A knowledge packet (*Kp*) is a reference to these concepts, their *C*, *U* and *L*'s (along with *mE*). For example, a *Kp* could represent a recycling technology (*En*) that is related to a cost (*Qn*), an energy consumption (*Rn*) in facility *An* where they are using a pricing mechanism *Tn*. Each *Kp* presents a picture of the state of the world *W* at time *t*. A knowledge supply chain is a film-reel of *Kp*'s.

Concepts and their modalities have to be built mainly (or at least, initially) based on bottom-up modeling of the social collective understandings of domain experts. Conceptualization is not a mechanical process (top-down and deterministic). It has to be built through consensus between users. Such consensus is relative, given that it is developed by a limited group of users. Ultimately, conceptualization cannot reach the "good enough" level from the first attempt. It has to go through iterative development to allow for adequate testing of its representativeness and re-balancing of the interests of the consenting group. This is not to say that top-down approach has no place in ontology development. As always, a balance has to be reached (for more, see El-Diraby 2012).

Integration and Merger Support

It is inevitable that many ontologies will co-exist in any domain. Even though many attempts have been conducted to support the integration of unstructured data, open questions remain regarding the mapping of different ontologies or linked data. At a fundamental level, to enhance our abilities for merging ontologies we need to re-consider the tree like nature of concepts and adopt a more network-like structure. Arranging concepts, relationships and even axioms into taxonomical hierarchies is limiting compared to a situation where we arrange them into a lattice. Use of formal concept analysis (FCA) can be very helpful in advancing the lattice or network-like style of concept topologies. FCA is a mathematical theory that models the concept of "concept". A concept is a unit that has an extent and intent. Concepts and attributes are represented in the form of lattice. Using lattice algebra, researchers have managed to provide semi-automated means to merge ontologies. Although representing ontologies in the form of lattices (on the basis of FCA) facilitates the task of ontology merging, some limitations require further consideration (El-Gohary and El-Diraby 2010).

In the interim, many web-based tools have used semantic systems to support some aspects of integrating unstructured data and/or ontologies. For example, the GIDS (Global Interlinked Data Store) technique distributes Linked Data across the network and then manages the network as a database (Braines et al. 2009). The SWEDER mechanism (Semantic Wrapping of Existing Data Sources with

Embedded Rules) makes available existing electronic data sources in a usable and semantically-rich format along with rules to facilitate integration between datasets (Braines et al. 2008). The POAF technique (Portable Ontology Aligned Fragments) aims to support alignment between existing ontology resources. These techniques (and many others) can be used to create interoperability between interlinked unstructured data sets based on semantic analysis (Kalfoglou et al. 2008).

Theoretically, and at a boarder scope, we can exploit topic maps for *ad hoc* match of unstructured data. They index multi-source information and can act as metadata (Pepper 2009). Topic maps can be used to match self-describing systems, which contain within themselves information about their use (Dailey et al. 2002). Further, probabilistic computational systems and prognostic normative reasoning (PNR) can help in integrating knowledge sources based on user context. PNR recognizes user profile and his/her needs, use various AI tools (such as user intention recognition, normative reasoning over a user's intention) to reason about the relevance of unstructured corpuses to these needs.

THE STRUCTURE OF ONTOLOGY

Internal vs. External Concepts

An ontology should not be limited to representing domain entities. It must provide views and means to describe entities outside its domain to allow for better chances to link to other knowledge sources in nearby/related domains—for example existing classification systems and IFC (see Ekholm, 1996).

Ad hoc Relationships

Relationships triangulate concepts and create mini-contexts, hence enhance the semantic representation of these concepts. By studying the in-going and outgoing relationships of a concept and considering other concepts in its network, we can define a semantic signature (context) of a concept. Tools such as GRDDL (Gleaning Resource Descriptions from Dialects of Languages) can be used for translating XHTML data into RDF format, whereby concepts are interlinked with relationships (see GRDDL 2013). Translating unstructured data into RFD create concept clusters, which makes such data valuable resource in identifying not only new concepts but also their semantic signature and can further facilitate the ad-hoc linking integration of such unstructured data sources into an ontology.

Nature of Axioms

First, axioms should attempt to describe what is not permissible rather than what is right. This opens doors for changes in our knowledge. Second, axioms should be modeled in a more flexible way. For example, the behaviour of agent *En* can be presented to express: *OEn* (when is it obligatory that *En*); *PEn* (when is it permitted that *En*); *FEn* (when is it forbidden that *En*). Thirdly, temporal subjective logic can help to contextualize axioms (Oren et al. 2012). Finally, we must assign relative weights to the axioms to allow for their violation if they are encountered by stronger axiom from another ontology or similar evidence form structured data.

Self-healing

Can the ontology adjust the concept names, relationship and axioms weights based on knowledge found in new or extended sources. Lessons learned from the self-healing nature of folksonomies are important here. Folksonomies are created collaboratively in a bottom-up manner. They evolve as circumstances change. Crowdsourcing and collaborative tagging and the resulting folksonomies can be used to create "tag clouds", which show the dominance of particular tags. Furthermore the user tagging activities can be used to identify things such as trending topics (Bao et al. 2009).

The use of opinion dynamics analysis tools and network theory can also be helpful in studying these trends. Opinion of each node i at each time step (t) can be modeled by a vector $\overrightarrow{O}_i(t)$, creating opinion matrix $[X_{(t)}]$. A weight matrix $[A]$ can be selected to reflect the effect of the other nodes based on their influence level (either found from analysis of the network structure, or by analysis of the discussions). A convincing mechanism (depicting the bound of confidence for each node, and its level of strength) can be used. The opinion profile of the project will be then updated at each step simply as: $[X_{(t+1)}] = [G] \cdot [X_{(0)}] + ([I - G])[A] \cdot [X_{(t)}]$ for $t \in T$. where $[X_{(0)}]$ refers to the initial opinion profile, and $[G]$ is the diagonal matrix of individuals' strengths.

ONTOLOGY STYLE

The style of ontology should also adopt to the needs of EMT. Ontology should embrace more bottom-up development approaches, distributed and collaborative authoring and support evolutionary changes to its contents. These techniques have been very helpful in the increasing impact of collaborative tagging and folksonomies in other fields. The following tools can help support these style changes.

More Linguistic Analysis, Less Formal Logic

Developers should emphasis the semantic nature of their ontologies to enhance expressivity. The use of Controlled Natural Language (CNL) interface capability can enhance access and use of extended sources to enrich the semantics of the ontology (Bao et al. 2009b). Further, recent advances in Natural Language Processing (NLP) and entity extraction also offer the potential for creating linked data from traditional unstructured textual data, such as blog entries or web pages—for example, OpenCalais from Thomson Reuters (see Calais 2013). For example,

Mounce et al. (2009) presents an example of semi automatic creation of ontology in the water field. They use the ontology learning tool Text2Onto.

In addition to traditional methods of semantic analysis, it is important to learn from research into human information way-finding on the web. This type of research investigates how people find, make sense, and track information online. Analyzing web navigation patterns can help discover relationships between knowledge constructs (bottom-up). If, on the one hand, we observe the network structure of knowledge sources that users visit, and, on the other hand, we study the semantic linkages between the main concepts in these sources, we can discover methods used by efficient information seekers in navigating unstructured data corpus. This can help in structuring and designing our ontology and the overall informatics system to be more human-friendly (West and Leskovec 2012).

Concept Typology

There are two main approaches for structuring concepts in an ontology (Hovy 2002). The first is the top-down approach of object-oriented models (which tracks its origins to Aristotle). It starts with a basic list of types that act as the root concepts and then grow the ontology around these concepts (just like a tree) by classifying them. While dominant, such approach has considerable drawbacks—mainly the difficulty of categorizing fuzzy concepts within the strict hierarchy of object trees. In contrast, philosophers (starting with Wittgenstein) and, recently mathematicians and computer scientist, have investigated what is called "prototype-oriented" approach for modeling. The basic paradigm in this approach is that instead of belonging to a strict hierarchy, objects belong to a "family of resemblance", where family members share a "degree" of resemblance (the notion of fuzzy function). The computer application of the prototype-oriented approach starts by identifying a set of objects that belong to a family. Elements of this family (objects) are stand-alone entities that do not have to contain similar attributes. To create a new object, an existing object (that best reflects the desired features of the new object) is cloned. Using modular operators, the cloned object is modified to reflect its essential characteristics. Late binding is used to ensure that methods defined earlier can be overridden to extend existing behavior (Taivalsaari, 1996).

Network Pattern Instead of a Top-down Structure

In light of the use of prototype oriented approach and emphasizing semantics, the way ontology entities (concepts, relationships, axioms) are arranged must also be changed. We have to accept a more networked view of concept instead of the traditional taxonomical (tree-like) view. This is closer to the format of FCA. It also makes ontology structure more similar to RDF-based linked data, which can add flexibility to ontological modeling given that RDF is more oriented towards localized representation rather than focusing on an overall structure or taxonomy of entities. Combining resolvable URIs and RDF can be very effective in navigating an extended network of linked sources of knowledge. This has been used in certain fields such a geographical data (Geonames), media and music business

(MusicBrainz) and collaborative wikis (DbPedia). The widespread adoption and refinement of such data sources will yield powerful points of common reference and disambiguation when referring to concepts and upon trying to manage co-references (Braines et al. 2010).

Further, initiatives such as the Semantically Interlinked Online Communities (SIOC) aim to provide a semantic basis for the integration of online social web community data. The Simple Knowledge Organisation System (SKOS) can support in the generation of taxonomical representation of extended sources of knowledge—allowing for better match to a base ontology. SPARQL, a standard Semantic Query Language, may be used to semantically search RDF-based data sources (Braines et al 2010).

ONTOLOGY DEVELOPMENT METHODOLOGY

Developing informatics ontologies as domain theories (not just computational models) and emphasising its epistemological aspects raise the interest in the process of ontology building itself. "Theory building can be described as "the purposeful process or recurring cycle by which coherent descriptions, explanations, and representations of observed or experienced phenomena are generated, verified, and refined (Lynham 2000)." Good theory building should result in two kinds of knowledge: outcome knowledge, usually in the form of explanative and predictive knowledge (i.e. the final model or theory), and process knowledge (increasing our understanding of how something works and what it means) (Dubin 1976). Theory building should also reflect two important qualities: rigor and relevance (Marsick 1990a), or what are also termed validity and utility (Van de Ven 1989). One of the most important tools to achieve these desired outputs is emphasize what is called "the logic-in-use" and the "reconstructed logic". This means that the development team should expose, track, analyze and critique its own underlying logical cognitive style in the development and application of the theory and by explicitly reconstructing, or making explicit the "logic-in-use". This enhances the development of a model or an ontology, and also enriches the wisdom and practices of doing so (Kapaln 1964).

Theory as a Tool

The polymorphic nature of today's knowledge mandates on developers of new ontologies to consider and test a variety of domain theories and/or ontologies to represent this knowledge. To increase the agility of ontology, we also have to break the traditional molds and emphasize an open mind regarding the existence of possible new worlds (or formats of knowledge). One important technique in this regard is to expose and question the hidden theory and/or assumptions that drive our traditional conceptualization of knowledge. Critiquing this underlying theory and experimenting with variations of it are among the most effective tools to discover a theory about the truth (see also Maxwell 1996; Johnson 1994).

Shields and Tajalli (2006) suggested the following tools to help expose and enrich theory as a tool practices:

- Formalizing *W*orking hypothesis (relevance: during the exploration phase of research): These are loosely formatted provisional, hypothesis meant to support advancing the investigation and the discovery of other critical facts. They should be carefully articulated at the start and questioned throughout the development exercise.
- Development of categories (relevance: during the description phase): clustering concepts into categories and/or universals.
- Practical *I*deal type (relevance: gauging): developing criteria to judge the effectiveness of inquiry and then to collect evidence to contrast the reality of the program against the criteria.
- Model of *O*perations *R*esearch (relevance: decision making): how to use the resulting model in decision making?
- Formal hypothesis (relevance: explanation and prediction): this micro-conceptual framework coincides with the "hypothetico-deductive" method of inquiry (see Kaplan 1964).

Triangulation

Systematic mixing and integration of a variety of opinions and points of views can help in assuring that ontology can meet different expectations and is flexible enough. Miles and Huberman (1994) distinguished five different types of triangulation in qualitative research:

- Triangulation by data source (data collected from different persons, or at different times);
- Triangulation by data type (e.g., combining quantitative and qualitative data);
- Triangulation by method (observation, interviews, documents, etc.);
- Triangulation by researcher; and
- Triangulation by theory (using different theories to explain results);

Scoping

Researchers should resist at all costs the idea that an ontology is universal. In fact, researchers should strive for a minimum ontological commitment. However, the famous minimalist approach (suggested by Gruber 1995) should only be dedicated to ontology scope (do not model outside your target domain). Minimalism (lack of depth or representativeness) is not acceptable within the structure of the ontology itself. Close attention to problem definition and requirements analysis procedures are keys for developing a clear and coherent scope. Competency questions have also a great role in making sure that the scope of the ontology is clear. Once the scope has been defined, the rigorous application of competency questions provides means to assure comprehensive coverage of concepts.

Iterative Development

As any model, as soon as an ontology is developed it helps enhance our knowledge of the domain including our assumptions. Consequently, it requires some modifications. Development steps and basic assumptions have to be revisited and calibrated frequently in an iterative manner before a stable and adequate ontology is reached. Repeated categorization and questioning the micro-conceptual framework of the ontology are of particular importance here. Emphasis on life cycle management (as suggested by Fernández-López and Gómez-Pérez 2002) and the triangulation of research methodology provide very helpful means in this regard.

SUMMARY

New trends in informatics systems aim to address knowledge management issue beyond the simple computational dimension and with acknowledgement of (focus on) the human and social aspects of knowledge. Mainly, integration of human-based (less mathematical) means of knowledge processing such as linguistics and communicative aspects, personal experiences and preferences, as well as contextual awareness into these systems. The traditional rule-based thinking must be replaced (or at least be complemented) by what can be called mining-based or statistic-based system. the traditional top-down approach for ontology development must be complemented with a bottom-up approach, where knowledge is discovered from open and extended sources.

The constructivism of EMT and the k-society denies the existence of an absolute ontology (reality is socially constructed) and advocates more temporal, relativistic and bottom-up models of knowledge. Consequently, the aim of modern ontology in this scope is to find a good-enough skeleton to describe the main concepts relating to construction knowledge that supports access and use of extended sources of knowledge. "*Can we suggest a network of categories to help represent, fairly adequately, most of our possible conceptual understanding of/about knowledge in the construction domain based on suitable perspectives of our perception of such knowledge without limiting the discovery or integration of ad hoc knowledge in unstructured/extended sources?*"

To elaborate: "*network of categories*" refer to the fact that the main claim/contribution of an ontology is its proposed categories of concepts. The word "*network*", in particular, emphasizes the role of relationships in the ontology and, more importantly, the resistance of the top-down structure. "*help represent*" is a declaration of the objective of ontology: knowledge representation not reasoning. And its role is not to exhaustively do that, it aims at just helping do that. "*fairly adequately*" and "*most*" recognize that due to the fuzzy nature of the concepts, ontology should not claim to be able to represent all concepts nor do so in the most accurate and definitive format. "*possible conceptual understanding*" refers to the rejection of realistic epistemology. In light of bottom-up nature of k-society, we have to embrace a constructivist approach (at the micro level) within a contemporary pragmatic umbrella (at the macro level). In this regards, ontology must accept possibilia; concepts have dual nature (that can change dynamically) with a probability of existence at different contexts (worlds). The word "*knowledge*" emphasizes the philosophical and conceptual nature of ontology. It refers to the target of representation: linguistically-rich, loosely-coupled wisdom gained by human beings—in contrast to information, which is synthetically structured data/fact or data itself (unstructured facts). "*suitable perspectives*" emphasizes "*good-enough*" solutions and the believe that knowledge is an interlinked possibilia that have relative formats based on the context of knowledge management exercise. "*perception*" refers to the fact that ontology should not claim a true correspondence to truth. "*without limiting*" is a call for minimalism in axioms to avoid restricting the discovery of new knowledge in extended sources. It also suggests that the aim of axioms should not be to exhaustively list facts. Rather, they should aim to list what is not acceptable. "*discovery or integration*" aims to shift the role of ontology to be a source of integration for unstructured knowledge rather than the sole encapsulation of what we know. "*?*" putting the whole scope in the form of a question emphasizes the fallibility of our models (in accordance with contemporary pragmatism).

References

Abramsky, S. and Jung, A. (1994). "Domain theory." *Handbook of Logic in Computer Science. III*, S., Abramsky, D. M., Gabbay, T. S. E., Maibaum, eds, Oxford University Press, Oxford, UK.

AIA—American Institute of Architects. (2013). *Integrated Project Delivery—A Working Definition, American Institute of Architects California Council*, Sacramento, CA.

Bao, J., Ding, L., Smart, P. R., Braines, D., and Jones, G. (2009). "Rule modeling using semantic mediawiki." *3rd Annual Conference of the International Technology Alliance (ACITA'09)*, Maryland, USA, 23–24.

Bao, J., Smart, P., Braines, D., and Shadbolt, N. (2009b). "A controlled natural language interface for semantic media wiki using the rabbit language." *Workshop on Controlled Natural Language (CNL'09)*, Marettimo Island, Italy, 8–10.

Bench-Capon, T. J. M. (1990). *Knowledge Representation; An Approach to Artificial Intelligence*, APIC series, No. 32, Academic Press, London, Toronto.

Borgo, S. (2007). "How formal ontology can help civil engineers." In: Teller, J., Lee, J., Roussey, C. (eds.) *Ontologies for Urban Development: Interfacing Urban Information Systems. Studies in Computational Intelligence*, vol. 61, pp. 143–156, Springer Verlag. ISBN 978-3-540-71975-

Braines, D., Jones, G., Smart, P. R., Bao, J., and Huynh, T. D. (2009). "GIDS: Global interlinked data store." *3rd Annual Conference of the International Technology Alliance (ACITA'09)*, Maryland, USA, 23–24.

Braines, D., Kalfoglou, Y., Smart, P. R., Shadbolt, N., and Bao, J. (2008). "A data-intensive lightweight semantic wrapper approach to aid information integration." *4th International Workshop on Contexts and Ontologies (C&O 2008)*, Patras, Greece.

Braines, D., Smart, P. R., and Bao, J. (2010). "Emergent Capabilities for Collaborative Teams in the Evolving Web Environment." *4th Annual Conference of the International Technology Alliance (ACITA'10)*, London, UK.

Calais (2013). ⟨http://www.opencalais.com/⟩ (?).

Chellas B. F. (1980). *Modal logic: an introduction*, Cambridge University Press, Cambridge, UK.

Chunara, R., Bouton, L., Ayers, J. W., and Brownstein, J. S. (2013). "Assessing the Online Social Environment for Surveillance of Obesity Prevalence." *PLoS ONE*, 8(4): e61373.

Connolly, D. ed (2013). "GRDDL-Gleaning Resource Descriptions from Dialects of Languages." ⟨http://www.w3.org/2004/01/rdxh/spec⟩ (accessed April 12, 2013).

Dailey, D. J., et al. (2002). "A self-describing data transfer model for ITS applications." *IEEE Trans. Intel. Trans. Sys.*, 3(4): 293–300.

Davenport, T. H. and Prusak, L. (1997). *Information Ecology*, Oxford University Press, Oxford, UK.

Davenport, T. and Prusak, L. (1998). *Working Knowledge*, Harvard Business School Press, Boston, MA

Davis, C., Nikolic, I., and Dijkema, G. (2010). "Infrastructure modeling 2.0," *Int. J. Crit. Infrastruct.*, 6(2), 168–186.

Diligenti, M., et al. (2000). "Focused Crawling Using Context Graphs." *26th Conf. on Very Large Databases*, Cairo, Egypt.

Dubin, R. (1976). "Theory building in applied areas." *Handbook of industrial and organizational psychology*, Dunnette, M. D., ed., Rand McNally, Chicago, IL., 17–39.

Dumontier, M. and Hoehndorf, R. (2010) "Realism for Scientific Ontologies." *6th International Conference on Formal Ontology in Information Systems*, Toronto, Canada.

Ekholm, A. and Fridqvist, S. (1996). "Modelling of user organisations, buildings and spaces for the design process." *Proc., CIB W78Workshop, Construction on the Information Highway*, Bled, Slovenia.

El-Diraby, T. E. (2011). "Infrastructure as a socio-technical chaotic system." *W78 Annual Workshop*, Sophia-Antipolis, France.

El-Diraby, T. E. (2012). "Epistemology of construction informatics." *J. Constr. Eng. Manage.*, ASCE, 138(1), 53–65.

El-Gohary, N. and El-Diraby, T. E. (2011). "Merging Architectural, Engineering and Construction (AEC) Ontologies." *J. Comput. Civil Eng.*, ASCE, in press.

Fernández-López, M. and Gómez-Pérez, A. (2002). "Overview and analysis of methodologies for building ontologies." *Knowl. Eng. Rev.*, 17(2), 129–156.

Floridi, L. (2004). "Open problems in the Philosophy of Information?" *Metaphilosophy*, 35 (4), 554–582.

Floridi, L. (2005). "Is semantic information meaningful data?" *Philos. Phenomen. Res.*, 70 (2), 351–370.

Granovetter, M. (1978). "Threshold models of collective behavior." *Am. J. Social.*, 83(6), 1420–1443.

Grant, R. M. (1996). "Toward a knowledge-based theory of the firm." *Strategic Manage. J.*, 17, 109–122.

Gruber, T. (1995). "Toward principles for the design of ontologies used for knowledge sharing." *Int. J. Hum. Comput. Stud.*, 43(5–6), 907–928.

Grüninger, M. and Fox, M. S. (1995). "Methodology for the design and evaluation of ontologies." *IJCAI95 Workshop on Basic Ontological Issues in Knowledge Sharing*, Montreal, Canada.

Harrison, W. and Ossher, H. (1993). "Subject-Oriented Programming: a critique of pure objects." *Proc., OOPSLA'93 Proceedings of the eighth annual conference on Object-oriented programming systems, languages, and applications*, ACM New York, NY, USA, 411–428.

Hastings, S., Ribeiro, M., Langella, S., Oster, S., Catalyurek, U., Pan, T., Huang, K., Ferreira, R., Saltz, J., and Kurc, T. (2005) "XML Database Support for Distributed Execution of Data-intensive Scientific Workflows." *ACM SIGMOD Record*, 34(3), 50–55.

Hendler, J. and Berners-Lee, T. (2010). "From the semantic web to social machines: A research challenge for AI on the world wide web." *Artif. Intel.*, 174(2), 156–161.

Hovy, E. (1998). "Combining and standardizing large-scale, practical ontologies for machine translation and other uses." *Proc., 1st Int. Conf. on Language Resources and Evaluation (LREC)*, Association for Computers and the Humanities (ACH), Granada, Spain, 535–542.

Johnson, R. B. (1994). "Qualitative research in education." *SRATE J.*, 4 (1), 3–7.

Kalfoglou, Y., Smart, P. R., Braines, D., and Shadbolt, N. (2008). "POAF: Portable ontology aligned fragments." *International Workshop on Ontologies: Reasoning and Modularity (WORM 2008)*, Tenerife, Spain.

Kaplan, A. (1964). *The Conduct of Inquiry: Methodology for Behavioural Science*, Chandler Publishing Co., Scranton, PA.

Kennedy, J., et al. (2001). *Swarm intelligence*, Morgan Kaufmann Pbl., San Francisco, CA.

Lynham, S. A. (2000). "Theory building in the human resource development profession." *Hum. Resour. Dev. Q.*, 11(2), 159–178.

Marsh, L. and Onof, C. (2007). "Stigmergic epistemology, stigmergic cognition" *Cognit. Syst. Res.*, 37(2).

Marsick, V. J. (1990a). "Altering the paradigm for theory building and research in human resource development." *Hum. Resour. Dev. Q.*, 1(1), 5–24.

Maxwell, J. A. (1996). *Qualitative research design*, Sage, Newbury Park, CA.

McDermott, D. V., Burstein, M. H., and Smith, D. R. (2001). "Overcoming Ontology Mismatches in Transactions with Self-Describing Service Agents." *The Emerging Semantic Web Conf.*, California, USA, 285–302.

Miles, M. B. and Huberman, A. M. (1994). *Qualitative Data Analysis*, Sage, Thousand Oaks, CA.

Morris, S. (2000). "Contagion." *Rev. econ. stud.*, 67, 57–78.

Mounce, S., Brewster, C., Ashley, R., and Hurley, L. (2009). "Knowledge management for more sustainable water systems." *Conference of the COST ACTION C21 – TOWNTOLOGY Urban Ontologies for an Improved Communication in Urban Civil Engineering Projects*, Liege, 39–49.

Oren, N., Bisdikian, C., and Chakraborty, S. (2012). "Temporal Subjective Logic", *6th Annual Conference of the International Technology Alliance*, Southampton, UK.

Pepper, S (2009). "The TAO of Topic Maps.", ⟨http://www.ontopia.net/topicmaps⟩ (?).

Quine, W. V. and Ullian, J. S. (1978). *The Web of Belief*, 2nd edition, Random House, New York, NY.

Qvortrup, L. (2006). *Knowledge Education and Learning—e-Learning in the Knowledge Society*, Forlaget Samfundsliteratur Roskilde Univ. Forlag, Frederiksberg, Denmark.

Rosato, V., Issacharoff, L., Tiriticco, F., Meloni, S., Porcellinis, S., and Setola, R. (2008). "Modelling interdependent infrastructures using interacting dynamical models." *Int. J. Crit. Infrastruct.*, 4(1), 63–79.

Roussey, C., Pinet, F., Kang, M. A., and Corcho, O. (2011). "An Introduction to Ontologies and Ontology Engineering." *Ontologies in Urban Development Projects, Advanced Information and Knowledge Processing 1*, G., Falquet et al., eds, Springer-Verlag, London, UK.

Santanchè, A., Medeiros, CB, and Pastorello, GZ (2007). "User-author centered multimedia building blocks" *Multimedia Syst.*, 12(2007), 403–421

Shields, P. M. and Tajalli, H. (2006). "Intermediate Theory: The Missing Link to Successful Student Scholarship." *J. Public Aff. Edu.*, 12(3), 313–334.

Smart, P., Huynh, T., Braines, D., Sycara, K., and Shadbolt, N. (2010). "Collective cognition: Exploring the dynamics of belief propagation and collective problem solving in multi-agent systems." *1st ITA Workshop on Network-Enabled Cognition: The Contribution of Social and Technological Networks to Human Cognition*, Maryland, USA.

Smart, P. R., Engelbrecht, P. C., Braines, D., Strub, M., and Giammanco, C. (2010b). "The network-extended mind." *Network Science for Military Coalition Operations: Information Extraction and Interaction*, Verma, D. ed., IGI Global, Hershey, Pennsylvania, USA, 191–236.

Smith, B. (2003). "Ontology." *The Blackwell guide to the philosophy of computing and information*, Floridi, L. ed, Wiley-Blackwell, New York, 153–166.

Sotoodeh, M. (2009). "Challenges in Semantic Interoperability in Emergency Management", *Proc., Advanced Information Systems Engineering, 21st International Conference, CAiSE 2009*, Amsterdam, The Netherlands.

Spink, A. and Park, M. (2004). "Information retrieval as multitasking: An exploratory framework." *ACM SIGIR 2004 Workshop on Information Retrieval in Context*, Sheffield, UK, 16–19.

Spyns, P. and Reinberger, M-L. (2005). "Lexically Evaluating Ontology Triples Generated Automatically from Texts." *Second European Semantic Web Conf.*

Stamatiadis, N. (2005). "Context-Sensitive Design." *J. Transp. Eng.*, ASCE, 131

Stehr, N. (2001). "Modern Societies as Knowledge Societies." *Handbook of Social Theory*, Ritzer, G. and Smart, B. eds., SAGE Publ., London, UK.

Strang, T. and Linnhoff-Popien, C. (2004) "A Context Modeling Survey." *Proc., the First International Workshop on Advanced Context Modelling, Reasoning and Management, Sixth International Conference on UbiComp2004.* Nottingham, England.

Stuckenschmidt, H. and Wache, H. (2000). "Context Modeling and Transformation for Semantic Interoperability." *Knowledge Representation Meets Databases*, 115–126.

Swindler, J. K. (1991). *Weaving: an analysis of the constitution of objects*, Rowman & Littlefield Publishers, Inc., Mayland, USA.

Taivalsaari, A. (1996). *Classes vs. prototypes: Some philosophical and historical observations*, Nokia Research Center, Helsinki, Finland.

Tapscott, D and Williams, A. D. (2007). *Wikinomics—How mass collaboration changes everything*, Portfolio/Penguin Publ., New York.

Tofani, A., Castorini, E., Palazzari, P., Usov, A., Beyel, C., Rome, E., and Servillo, P. (2010). "An ontological approach to simulate critical infrastructures." *J. Comput. Sci. Eng.*, 1(4), 221–228.

Van deVen, A. H. (1989). "Nothing is quite so practical as a good theory." *Acad. Manage. Rev.*, 14(4), 486–489.

Van Herzele, A. H. (2004). "Local knowledge in action valuing nonprofessional reasoning in the planning." *J. Plan. Educ. Res*, 24(2), 197–212.

Visser, P. R. S. and Bench-Capon, T. J. M. (1997). "A Comparison of Four Ontologies for the Design of Legal Knowledge Systems." *Artif. Intell. Law*, 6(1), 27–57.

Vold, K. (2011). "A Defense of the Extended Mind Thesis", *Prometheus*, 4(1)

von Hippel, E. (2005). *Democratizing Innovation*, MIT Press, Cambridge, MA.

West, R. and Leskovec, J. (2012). "Human wayfinding in information networks." *22nd International World Wide Web Conference*, Lyon, France.

Zhai, Z. (2012). "Citizen engineering: methods for 'crowdsourcing' highly trustworthy results." *45th Intern'l Conf. on system science*, Honolulu, HI.

CHAPTER 6

The Potential of Ontology-Based Serious Game Design for the AEC Domain

Kristian Schatz*
Uwe Rüppel*

Abstract: *The term "Serious Gaming" or "Gaming with a purpose" describes the combination of game concepts and methods with other information technologies in "serious" fields of applications. In the context of the Architecture, Engineering and Construction (AEC) domain Serious Games have the potential to achieve additional value beyond pure entertainment. During the game design process the thematic background (in the presented case the AEC background) of a Serious Game is of particular importance. This requires domain-specific knowledge, usually represented by domain-experts (civil engineers and architects) as part of the game development team. A challenging task of the design process is the creation of game objects to model the game world. In the presented case these game objects are based on common AEC objects (like wall, floor, window, door etc.) with some additional information about their role in the game scenario like interactive behavior or different states during the game timeline. To enhance the AEC objects information about the implicit meaning of each object is required. This includes a time consuming reasoning and decision making process based on the experts experience in the application domain. However, it is assumed that this currently limits the application of Serious Gaming in the AEC domain. To solve this problem the presented research approach tries to retrieve most of the required knowledge automatically form a knowledge base (KB). This KB consists of digital content, which is already created during the Building Information Modeling (BIM) process. The challenge here is to displace the interpretation of information from the expert into a computer system. This requires a formalization of knowledge that could be derived from existing data and captured by an ontology, including a set of concepts and relationships between those concepts within the application domain.*

---

*Institute of Numerical Methods and Informatics in Civil Engineering, Department of Civil and Environmental Engineering, Technische Universität Darmstadt, Germany email: schatz@iib.tu-darmstadt.de, rueppel@iib.tu-darmstadt.de

The present chapter introduces an ontology-based approach to support Serious Game design by enabling automatic creation of game design documents and game content using BIM data as a KB. For creating the KB for the presented use case, single AEC objects were enriched with additional information (e.g. events for interaction, animations, artwork, textures) about their role in the game with semantic technologies. This approach has the potential, that Serious Gaming becomes a common engineering method for daily practice, because data that are already incurred in the BIM process can be integrated and reused with minor effort.

INTRODUCTION

In the context of the Architecture, Engineering and Construction (AEC) domain Serious Games can be used for digital-game based learning, training and simulation, interactive assessments and storytelling, data collection, exploration or theory testing only to address a few. This chapter describes an approach how semantic technologies and especially ontologies could support the Serious Game design process in this context to add behavior modeling into the Building Information Modeling (BIM) process. Table 6-1 shows an outline how this chapter is structured. The first three sections are written as an introduction to Serious Gaming. In the end, the question is how to bring together BIM and game concepts. Semantic technologies are seen as an answer to solve existing problems to enhance BIM data about game playing behavior. The next two sections are of particular interest, because they address the concept of using semantic technologies (ontologies and semantic annotation) and its implementation is described in detail. This is followed by a use case before concluding with a summary and outlook.

PRELIMINARY THOUGHTS ON SERIOUS GAMING

Before going more into detail about these use cases of Serious Games in the AEC domain this section presents some general ideas and thoughts about games.

The central task, people have to carry out for serious gaming, is playing. How can "playing" be serious? It is fun, entertainment etc. and on the first view it has nothing to do with our daily work. Harteveld highlights this contrast. He writes: *"Adults do not play. . . . Adults should work, be serious, act responsible, and take care of others. . . . The thing is, most adults actually do play! They play in their work and in their private live. Adults only do not (dare) to call it that way."* (Harteveld, 2011, p. 175). Huizinga, a Dutch historian and cultural philosopher, has argued that "gaming" or "playing" is of prime importance for the development of our culture and he sees the people as player/gamer, as "Homo Ludens" (Huizinga, 1938). If we take a look back into the history of mankind since the early beginning, we find that many innovations and discoveries go back to a ludic drive or play instinct. Early humans were just "playing" around with some woods and stones

Table 6-1. Chapter outline

<table>
<tr><td rowspan="13">Chapter</td><td rowspan="2">Introduction</td><td colspan="2">1 Preliminary thoughts on Serious Gaming</td></tr>
<tr><td colspan="2">2 Use Cases for Serious Gaming approaches in the AEC domain</td></tr>
<tr><td rowspan="4">Analysis</td><td rowspan="4">3 Building Information Gaming</td><td>Game development workflow</td></tr>
<tr><td>BIM domain</td></tr>
<tr><td>3D-modeling and animation domain</td></tr>
<tr><td>Game domain</td></tr>
<tr><td rowspan="2">Concept</td><td rowspan="2">4 Semantic technologies for Building Information Gaming</td><td>Ontology-based Building Information Gaming</td></tr>
<tr><td>Semantic annotation of building elements</td></tr>
<tr><td rowspan="2">Imple-mentation</td><td rowspan="2">5 Building Information Gaming environment</td><td>Components of the environment</td></tr>
<tr><td>Graphical User Interface (GUI)</td></tr>
<tr><td>Use Case</td><td colspan="2">6 Serious Human Rescue Game design model</td></tr>
<tr><td>Conclusion and Outlook</td><td colspan="2">7 Conclusions and Outlook</td></tr>
</table>

and they noted that this could be useful tools for they daily work like hunting or farming. They refined their tools through trial and error and they trained to deal with them in competitions, which have essential elements of a game. However, it could be argued that "playing" or "gaming" is deeply rooted in human nature and it cannot be understood as a self-contained activity for fun after work. Rather, our daily life is interwoven with playful elements and actions. The question is: How could this potential be used to increase our productivity? Prensky, an expert on the connection between learning and technology, shares some ideas and thoughts in his publications to find an answer to the question: How digital technologies can enhance people's minds and lead to greater wisdom?

Prensky points out, that in the 21st century most people will be "Digital Natives". According to Prensky a "Digital Native" is a *"native speaker" of the digital language of computers, video games, and the Internet* (Prensky, 2011, p. 67). Prensky lists some reasons why in his opinion peoples wisdom is limited today (Prensky, 2011, p. 206):

- *People make decisions based on only portion of the available data*
- *People make assumptions, often inaccurate, about the thoughts or intention of others.*

- *People depend on educated guessing and verification (the traditional scientific method) to find new answers.*
- *People are limited in their ability to predict the future and construct what-if scenarios.*
- *People cannot deal well with complexity beyond a certain point.*
- *People cannot see, hear, touch, feel, or smell beyond the range of their senses.*
- *People find it difficult to hold multiple perspectives simultaneously.*
- *People have difficulty separating emotional responses from rational conclusion.*
- *People forget.*

To overcome these limitations, Prensky argues that in the future digital tools will extend and enhance the people's cognitive capabilities and that this progress will lead in "Digital Wisdom". Prensky describes a few examples of how he sees this enhancement in the future: One example deals with pairing human intelligence with digital simulation to conduct deeper analyses. According to Prensky the human ability to interpret and evaluate the models underlying the simulations plays a large role in using them wisely. Prensky claims that in the future, *more sophisticated simulation algorithms will allow humans to exercise their imaginative capacity in ever-more complex what-if constructions, allowing for more thorough exploration of possibilities and wiser decisions* (Prensky, 2011, p. 208). According to Prensky modern simulation games have the potential to support this. In another example Prensky asserts that digital tools have the potential to enhance the access to alternative perspectives, especially when some things are too small, too large, too fast, too abstract, too dangerous or outside the range of humans unaided senses. Prensky points out, that exploring these things through digital enhancements will help to get a better understanding of these things and also expand the ability to analyze things from more than one perspective. As an example Prensky suggests using interactive three-dimensional simulation (games) (Prensky, 2011). However, according to Prensky, digital wisdom arises from the combination of the mind and digital tools.

Jane McGonigal goes one step further. She not only wants to enhance human capabilities, she has the vision to fix the reality with games (McGonigal, 2012). According to her, in today's society computer and video games are fulfilling genuine human needs, because so many people are choosing to spend much time in game worlds. McGonigal argues that one reason is, that Games are providing rewards, the real world currently not. She suggests using everything we know about game design to fix what's wrong with reality to solve real-world problems and to make the real life better (McGonigal, 2012).

Serious Games can be seen as such digital tools. Admittedly, these are very visionary thoughts. Summarized it can be said that computer games are seen as a trend that will affect many aspects of daily life in the future. However, computer games are already used today in many productive areas and use-cases. The question is: In which areas of the AEC domain such tools could be used? The next section tries to give an answer on this and gives a brief introduction in Serious Gaming and its possible value creation by focusing on use-cases for Serious Games in the AEC-domain.

SERIOUS GAMING USE CASES IN THE AEC DOMAIN

The approach of Serious Gaming combines fun methods and concepts as well as game technology with other information and communication technologies (e.g., sensors, computer graphics, multimedia, artificial technology) and sciences (e.g., computer science, design, psychology, pedagogy) in "serious fields of applications".

Serious Games are designed to achieve additional values beyond the pure entertainment use. But what can these additional values be in the AEC domain? With the ideas of Prensky for digital enhancements in mind, Figure 6-1 shows some conceivable use-cases where a Serious Gaming approach might be able to add additional values.

The use cases shown in Figure 6-1 can also be compared with the values, Harteveld listed in his description of the world of meaning of his Triadic Game Design (TGD) approach. According to Harteveld, the approach of TGD involves a

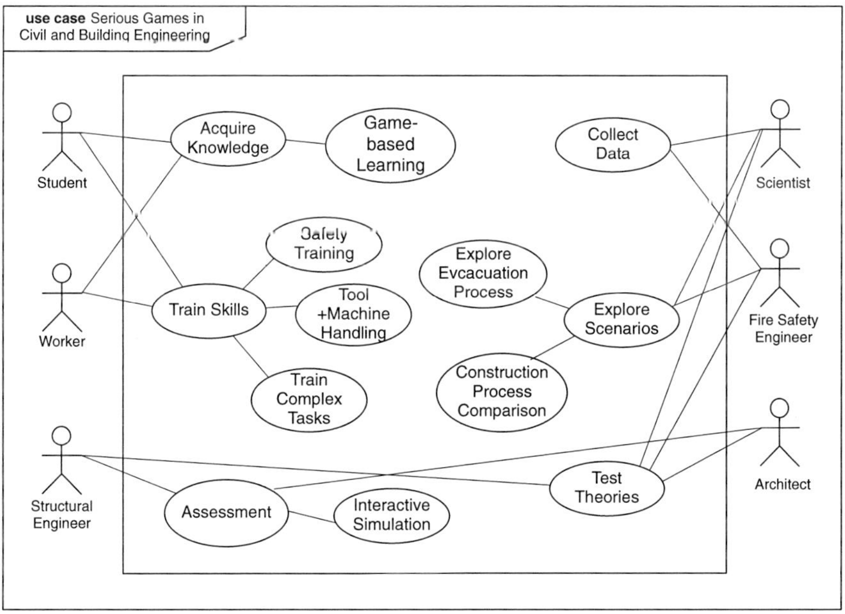

Figure 6-1. Potential Use Cases for Serious Games in the AEC domain

triad consisting of the interdependent worlds of reality, meaning and play that have to be balanced out during the game design process (Harteveld, 2011).

The *world of meaning* focuses on the creation of an effect, which is useful in the real world. The main question to answer is: What is the additional value to achieve?

For the AEC community these useful effects for example could be:

- *Knowledge:* Civil-engineering students might acquire knowledge by getting immersed into a specific, building related game scenario. Game-based learning can be seen as a method to get an understanding of what happened during that scenario.
- *Train Skills:* Workers or craftsmen could play a game for safety training, to learn how to handle special tools or to train complex tasks of a construction process. This could be useful to apply what is learned beyond the context of the game. In this case a serious game is used for the acquisition or exercise of different skills
- *Assessment:* Structural engineers could assess their simulation results in an interactive way. They can judge systematically different structural designs in a safe game environment.
- *Data Collection:* Data about a specific user behavior in a building could be collected with a game, which is useful for other purposes, e.g., for optimization of guiding systems.
- *Exploration:* Construct a scenario or simply brainstorm about all kinds of possibilities without having a clear idea. Researchers step into or create some environment and just see what happens. In all of these instances people attempt to "explore" a certain topic.
- *Theory Testing:* In contrast to exploration, games could also be used when people have a clear idea upfront. In this case they would like to "test" if their preconceived notions, based on theories, assumptions, or anything else, hold ground. The game becomes a sort of "laboratory experiment" in which researchers "play" with variables.

According to Hartevelds TGD approach the *world of reality* is responsible for how the game is connected with its thematic background. At this point in the game design process, domain-specific knowledge from the field of AEC is required. One possibility is to retrieve this knowledge from the Building Information Modeling (BIM) process by using BIM data as a knowledge base (KB). During the BIM process the focus is on data management of product model for describing the building with geometric and semantic expressions. Building objects (like walls, doors, windows) are represented as 3D objects with additional semantic information. BIM software tools can provide access to the information, visualization, and simulation capabilities of the digital building model.

These two worlds, the *worlds of meaning* and *reality*, now have to be balanced out with the *world of play*. Harteveld claims that the aspects of the *world of play* concern the goal of the game, its gameplay, the game world and the technology used. Key questions to answer are:

- What does the Player do? What is the goal of the game?
- What actions can the player perform to reach the goal?
- What are the challenges? The obstacles the player must overcome?
- What are the elements of the game world?

These questions are later referenced as "Gameplay Questions".

The presented approach is based on the experiences that have been made in the context of the project "Serious Human Rescue Game (SHRG)" (2009–2011) at Technische Universität Darmstadt (Rüppel, Schatz 2011). In this project the Institute of Numerical Methods and Informatics in Civil Engineering (IIB) and the Institute of Psychology are working together to research for "human factors" in the evacuation process. In this case the additional values are data collection, scenario exploration and knowledge acquisition about human decision-making during the evacuation process. The challenges in designing a Serious Game for this purpose are firstly, to model the game scenario according to the building design and secondly, to enhance the model with an authentic simulation of the emergency scenario (e.g., fire, smoke, explosion). To make sure that the model is valid – underlining the "serious" of serious gaming – it is essential that the underlying fire and smoke simulation model is comparable to state of the art fire safety engineering simulations.

The objectives of the SHRG will consider a fire scenario as one possible scenario of a building disaster. In this virtual fire scenario a player controls a pawn with a special amount of vitality. The main objective for the player is to guide his pawn through the virtual game scenario to a secure area before his vitality decreases until zero. Optional objectives could be, that he safes as much vitality as possible or that he has to find the quickest way to a secure area. So it is required, that the information that influence the vitality and the sense of orientation of a person in case of fire could be provided in the game scenario. Another requirement is, that it should be possible, that the pawn can move through the virtual building and handle virtual objects in the same way as a real person through a real building (e.g., pass doors if they are not locked or handle equipment)

From the experience in the context of the SHRG project the next section describes one possible workflow to create game content (the game elements) based on existing digital building data using existing file interchange formats.

BUILDING INFORMATION GAMING

Game Development Workflow

Figure 6-2 shows a flow-chart that describes the workflow for creating game content starting with the digital building model by using data exchange between different software tools. In the following the single tasks are described by a simple

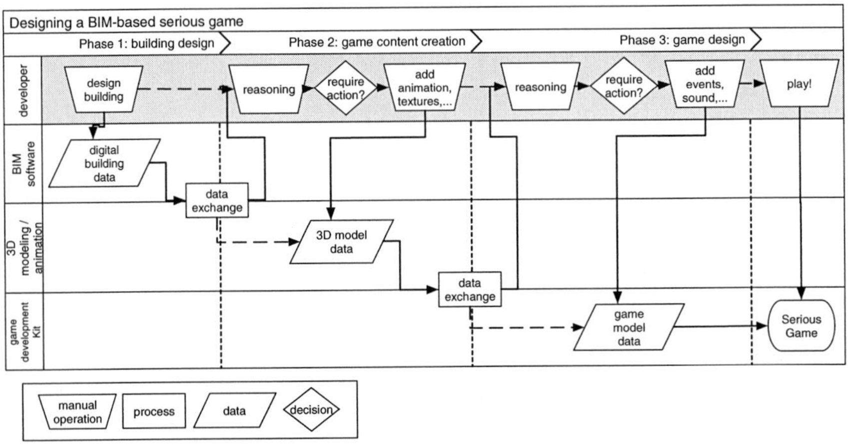

Figure 6-2. Designing a BIM-based Serious Game

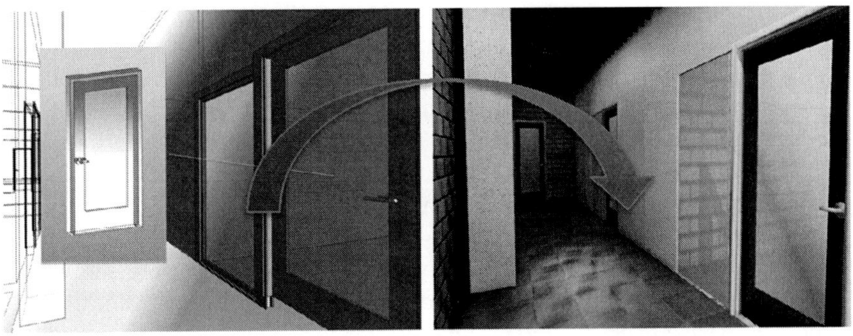

Figure 6-3. Simple door visualization: BIM software (left) and game scenario (right)

example. The example consists of a common building element: A *door* that is placed as an opening in a wall (Figure 6-3).

- *Phase 1:* Building design with common BIM-Software (e.g. Autodesk Revit Architecture). During this phase the developer uses common BIM software for the design task. The two elements wall and door are added to the digital building model. After this phase is completed, the geometrical building data should exported to an exchange-format, which can be used for import in 3D modeling and animation software for game content creation.
- *Phase 2:* This phase starts with importing the geometrical data into 3D modeling and animation software. The developer adds materials and animation to the 3D models of the building elements. In case of the door example, animations for the behavior of a door, in this case for the open and closing process must be added. After the developer finished this manual operation, the data must be exported again to an exchange-format, which could be imported into the game development environment as game content.

- *Phase 3:* Import game content. The game content created by the developer out of the digital building model in phase 2 could be imported into the game development environment.

In every phase a lot of work is needed each time to manually develop the game content based on the imported BIM objects. For every geometrical representation of a building element in the 3D modeling environment, the developer must make a decision if action is required to add additional information. Questions about each object are to answer. These "Object questions" are for example:

- Is the object static or dynamic? Destructible or indestructible? Visible or invisible?
- Which activity is associated with the object?
- Does the object can have different states? What states?
- Which characteristics of the object affect other objects? On what way? On what other objects?
- Is the object localized or freely movable?
- Is the object part of another object? Composed of other objects?
- Is the object source or drain for resources or other objects? Which resources (e.g. time, money, energy, raw materials, scores) and in what quantity?

To answer these questions, reasoning about the implicit meaning of each object in the game context is required. The product of the described workflow could be described with the term "*Building Information Game (BIG)*". BIG and the associated processes are defined as follows:

- A *Building Information Game (BIG)* is a digital model of a computer game, which contains information about relationships of game objects to digital building objects of a Building Information Model. A BIG is the product of the synthesis of a digital building model and a computer game model.
- *Building Information Game Design (BIG Design)* summarizes all processes for the specification of a *Building Information Game.*
- *Building Information Gaming (BIGing)* describes the application (playing) of a *Building Information Game.*

However, these manual operations of the BIG design process are typically structured in operations containing a number of recurring similar sub-operations. From the experience gained in the SHRG project this is a very time consuming phase.

The goal of the presented research is to limit the number of manual operations using semantic technologies for developing a knowledge base in order to provide answers to most of the questions above. For this approach it is assumed that all the building-related information exists in structured way and a standardized file format (here the IFC format) and that the BIM software supports a common 3D-model file export format (here the FBX format). Another requirement is, that the relationship between digital building model and 3D-model could

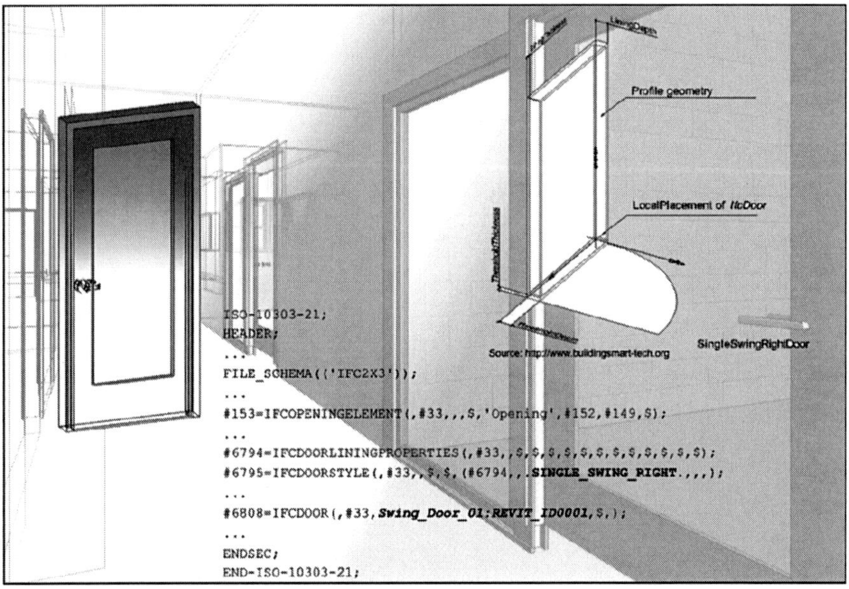

Figure 6-4. IFC data stored in IFC file (excerpt)

be easily identified (here on a object-ID base). At least, a clear description of the additional meaning and role of the objects for gameplay must be provided by a conceptual view. To point out what this "additional meaning" in the game context could be, the next sections tries to give a brief introduction to this domains.

BIM Domain

BIM software tools provide various export formats for different application domains. Within the BIM domain the Industry Foundation Classes (IFC) is a common exchange technology. Although possibilities for the description of geometric building objects are included in the IFC classes, possibilities for describing dynamic behavior of objects, such as animations for opening a door, are missed. Figure 6-4 shows the excerpt of the IFC file of the simple door example. The first highlighted expression is the information about the opening operation of the door (this must be interpreted for adding the correct interactive behavior). The second highlighted expression is the element name (in this case given by the BIM software Autodesk Revit Architecture).

3D-Modeling and Animation Somain

Within the domain of 3D digital content creation applications, a wide range of software tools supports the FBX technology. The FBX technology includes a FBX file format for 3D scenes and the FBX Software Development Kit (Autodesk, 2012). The data stored in the FBX files (*.fbx) can be manipulated over the FBX Software Development Kit provided by Autodesk and can be saved in an ASCII

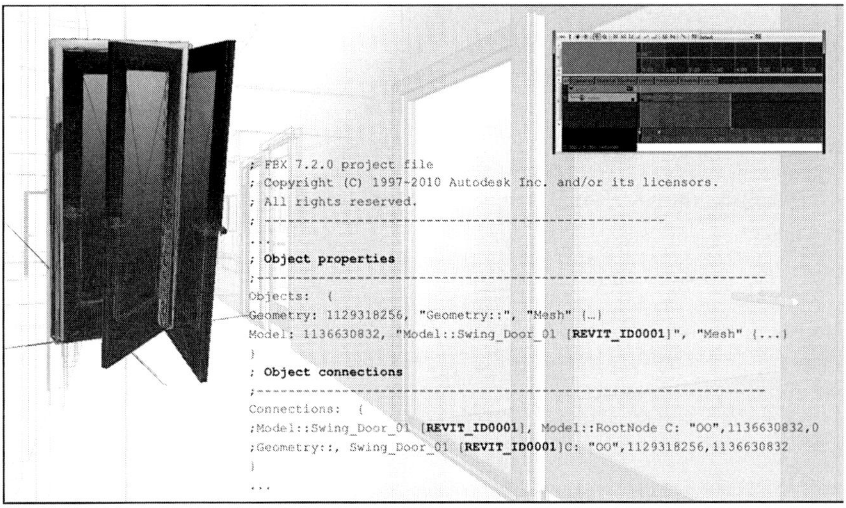

Figure 6-5. FBX data for door animation stored in FBX ASCII file (excerpt)

format. The FBX SDK provides to access, create, or modify the elements of a 3D scene (*KFbxScene*). A 3D scene is organized as a scene graph with a hierarchy of nodes (*KFbxNode*). The geometry of each building element is represented by *KFbxMesh* and attached to a *KFbxNode*. Figure 6-5 shows the excerpt of the FBX file of the simple door example.

In the highlighted section *Object properties*, the description of each object is stored. In the section *Object connection* the connections of the objects in the hierarchical order of the scene graph is stored. The third expression, which is highlighted three times, is again the element name. In this case, the source application for both files (IFC and FBX) was the BIM software Autodesk Revit Architecture, so the elements could be mapped over the given name.

Game Domain

The SHRG was realized with the Unreal Development Kit (UDK), a free edition of the Unreal Engine 3 (UDK, 2013). Bundled with the UDK comes the Unreal Editor, a modeling software for the creation of three-dimensional game scenarios. To get an understanding how the game data is structured it is important to make clear what happens during the game loop. The game loop is an interaction cycle between a player and a computer system and describes the process of updating the virtual world, if a game object has changed its state as a result of a user-input and of generating a new image (frame) that shows the actual view of the virtual world. The user-input is provided by Human Interface Devices (HID) like joystick or gamepad. The updated scene can be viewed at various kinds of displays. How many times per second a new frame can be generated is measured as frame rate with the unit frames per seconds (fps). This frame rate is crucial for playing the computer game.

Claypool (2007) showed that the frame rate has a direct influence on the playability of a game and that the frame rate is critical for adequate game performance. It was found out that frame rates lower than 7 fps are almost unplayable and that performance benefits for gamers could be measured up through 60 fps. This means that for a fluent gameplay, the game loop must be repeated at least 60 times per second.

For entertaining computer games the focus is more on special effects, for example fire and smoke have to look realistic and spectacular. To be able to calculate such kind of simulations in real time during the game loop, generally strong simplifications are made in favor of performance and not conducted entirely correct physics calculations. In some games, for example, fire is only animated with no spread and smoke emission. The hardware and software setup are responsible for processing the game loop inside the gaming engine.

According to Gregory, a game engine is "*arguably a data driven architecture what differentiates a game engine from a piece of software that is a game but not an engine. When a game contains hard-coded logic or game rules or employs special-case code to render specific types of game objects, it becomes difficult or impossible to reuse that software to make a different game. We should probably reserve the term "game engine" for software that is extensible and can be used as the foundation for many different games without major modification*" (Gregory 2009). In general, a game engine is modular and consists of several components. The virtual reality in which the game takes place will be referred to as the game world or game scenario. The virtual game world itself is composed of many different world elements. According to Gregory this world elements can be classified into two categories: static and dynamic elements. Static elements can generally be described as simple 3D-objects consist of triangular faces defined by vertices (generally called mesh) and some render-appearance definitions for the material (texture, reflections) to setup the visual behavior of the 3D-object. The dynamic elements behave normally interactive and their behavior has an influence on the game and the game play. For the simple door example the door as dynamic game object became interactive by a *Trigger* (see Figure 6-6).

The term *Trigger* is used in many game development kits for an invisible game object, which is connected to another game object to control its dynamic behavior. In the given example the trigger, which is connected to the door observes the game world for touch events. If a player-controlled avatar gets into the area (touched) of the door, the trigger recognizes this and will play the opening animation of the door. If the avatar leaves the area (untouched), the animation will be played in the reverse mode for closing the door.

The gameplay is concentrated within the dynamic elements. In terms of computing hardware resources needed for processing the game loop, dynamic elements are more expensive than static elements. The higher the ratio of dynamic to static elements is, the more "alive" the game world will appear to the player (Gregory, 2009).

But with this division into static and dynamic elements still no statement can be made about the meaning of each item for the gameplay and what kind of event

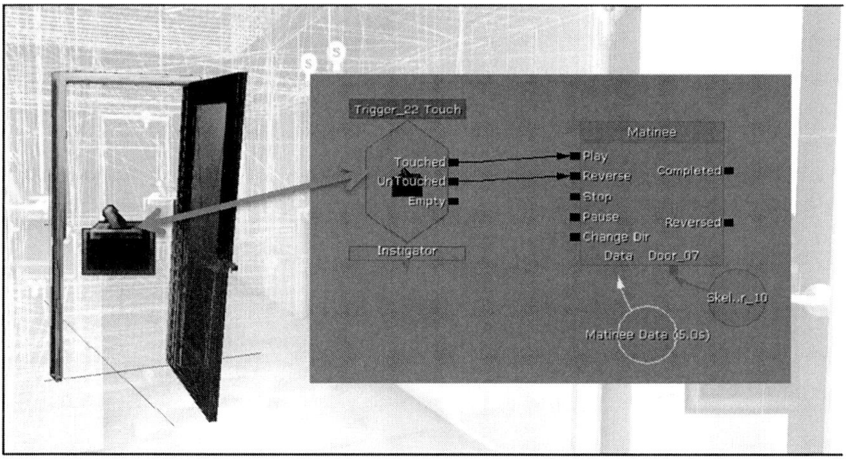

Figure 6-6. Dynamic game object door controlled by a trigger

Table 6-2. Gameplay Bricks (Adapted from Djaouti et al. 2007)

	Game Bricks
Avoid	Asks the player to avoid elements/ traps/ opponents.
Match	Asks the player to match or to keep one or several elements in a particular state.
Destroy	Asks the player to destroy/collect/catch elements or opponents.
Block	Asks the player to block elements or opponents.
	Play Bricks
Create	Allows the player to assemble, build or create elements.
Manage	Lets the player manage various resources in order to perform actions.
Move	Lets the player drive/pilot/displace an element or a character.
Random	Lets luck attributes values to the player.
Select	Lets the player select an in-game element by any input device (mouse, keyboard, gamepad . . .).
Shoot	Lets the player throw or shoot elements.
Write	Lets the player inputs an alphanumerical value.

or action sequences the *Trigger* must observe. Djaouti et al. introduced the term "Gameplay bricks", which describes "*a kind of "fundamental elements" whose different combinations seem to be able to cover the gameplay of computer games*" (Djaouti et al. 2007). Djaouti et al. have identified eleven "Gameplay Bricks" (see Table 6-2) by trying to find the most recurrent gameplay rules patterns. They define gameplay as: "*the association of "Game rules", stating a goal to reach, with "Play rules", defining means and constraints to reach this goal*". In addition, they introduced a third category of rules, the "World rules". To sum up Djaouti et al. propose an extended topology to categorize the gameplay rules patterns (Djaouti et al. 2007):

- "*Play rules*", acting on game elements according to player's input.
- "*Game rules*", watching the state of game elements in order to judge player's performance.
- "*World rules*", running the simulation that allows the virtual game world to "come to life" (e.g. A.I. or Physics).

According to Djaouti et al these three kinds of rules aren't in direct relation, they are "communicating" through the game elements.

Based on the gameplay rules categories, Djaouti et al. split the "Gameplay Bricks" into "Play Bricks", which are linked to user-input and "Game Bricks", which are linked to the goal.

During their studies Djaouti et al. have noticed that some pairs of "Game Bricks" and "Play Bricks" were found very often and that it seems that they are related to the core challenges proposed by games. They named this pair bricks "Metabricks". For the SHRG project the core challenge can be described in natural language with the sentence (1):

The player moves his pawn in order to avoid fire and smoke. (1)

This sentence can be divided into two clauses. The first clause "*The player moves his pawn*" defines the relationship between "*Player*" and the game element "*Pawn*" with the "move" action, while the second clause "*in order to avoid fire and smoke*" defines the goal to reach. The core challenge of the SHRG could be translated in the Metabrick "*Evacuate*" with the statement (2):

$$\boldsymbol{EVACUATE\ (Meta\,Brick) = MOVE\,(Play\,Brick) + AVOID\,(Game\,Brick)} \tag{2}$$

Some meaning is explicit included in this sentence while other meaning is implied. However, the description of the meaning and functional aspects of building elements and how they interact with users (in reality or in a virtual computer game) are purely syntactic nature (manual operations required to enhance the imported 3D-model for interactions). This results in a semantic gap between the syntactic description of a component and the underlying meaning. Therefore, searching, finding, selecting and combining gameplay elements is a complex task, especially with regard to storytelling and level-design. To solve this problem the aim of the proposed approach is to make the implicit meaning of machine-recognizable with the help of semantic technologies.

SEMANTIC TECHNOLOGIES FOR BUILDING INFORMATION GAMING

This section describes the development of an approach to support the BIG design process by enabling automatic creation of game design documents and game content (the game elements) using semantic technologies to link the three domains BIM, 3D-Modeling and Gaming. The workflow, described in the previous section, requires a lot of manual operations that were traditionally regarded as an intellectual task of the developer. This includes reasoning and decision making based on the developers experience in the application domain. The challenge here is to develop a knowledge base (KB) to displace the interpretation of information from the developer into a computer system. A KB typically consists of two main components. The first component describes the main concepts, properties and relationships of the domain. This component is called the "terminological component" or TBox. The second component stores facts about individuals associated with the TBox-concepts, this component is called "assertion component" or ABox.

The next two sections describe the development of the ontology for the TBox followed by using semantic annotation to define individuals for the ABox.

Ontology-Based Building Information Gaming

The knowledge could be captured by ontology as a set of concepts and relationships between those concepts within the application domain. A logical formalism for the development of ontologies is provided by Description logic (DL), a family of formal knowledge representation languages. The Web Ontology Language (OWL), which was chosen to express the ontology in the presented approach, is part of this family (W3C, 2009). Common elements to describe ontology are concepts, instances and relations between them. Usually relations in ontology are represented through the hierarchy of classes. An introduction to concepts and an overview of ontologies can be found in Stuckenschmidt (2011).

The present ontology was developed following the steps proposed by Noy and McGuinnes (Noy and McGuinness 2000). The competency questions the ontology should be able to answer are the "Gameplay questions" and "Object questions" introduced earlier in this chapter.

In the search for reusable ontologies for the game domain the Game Ontology Project was investigated (GOP 2005). But it was found out that there are logical faults and inconsistencies in the classes-subclasses hierarchy of the described concepts that makes a reuse inappropriate for the described application.

However, it was decided to develop a new ontology based on the Gameplay-Brick pattern described in (2). For this ontology development a top-down approach, starting with the definition of the most general concepts, was chosen. The TBox ontology is build and maintained with Protégé-OWL, a free and open source ontology editor. Protégé ontologies can be exported into a variety of formats including OWL. In Protégé, concepts are represented by classes and could be organized into a superclass-subclass hierarchy (taxonomy). Classes are interpreted as sets that contain individuals. Individuals are also known as instances and can be referred to as being 'instances of classes'. At least relations on instances can be described via properties. An introduction to building ontologies with Protégé can be found in Horridge (2011).

The main idea of the proposed approach is to use the ontology on the one hand data integration and on the other hand to enhance the BIM objects. Ontology-based data integration is based on the idea of decoupling information semantics from data storage (Calvanese 2011). This approach requires to develop some kind of global view on the data that integrates the domains of interest, in this case the domains BIM and 3D modeling for game content creation. The global view acts as a conceptual layer that could be accessed for query answering. The data layer is hidden for the requesting client. For the presented research, the data layer must include the digital building data as well as 3D model data.

For both data structures (IFC and FBX) a corresponding taxonomy in the sense of a connection layer was defined in the developed ontology. Now the ontology includes the main concepts like *ifcWall* or *ifcDoor*, which are connected with the corresponding concepts for the representations of those objects in a 3D scene like *KFbxNode* and *KFbxMesh*. Figure 6-7 shows an extract of the TBox ontology including concepts and relationships of the BIM and the Game domain.

If a more detailed view on the data is required the concept of the TBox includes two programming interfaces. The first interface is the IFC Java Toolkit to query IFC files (IFC Tools Project 2013), the second to the FBX Software Development Kit to query FBX files (Autodesk 2012). An *ifcElement* can be

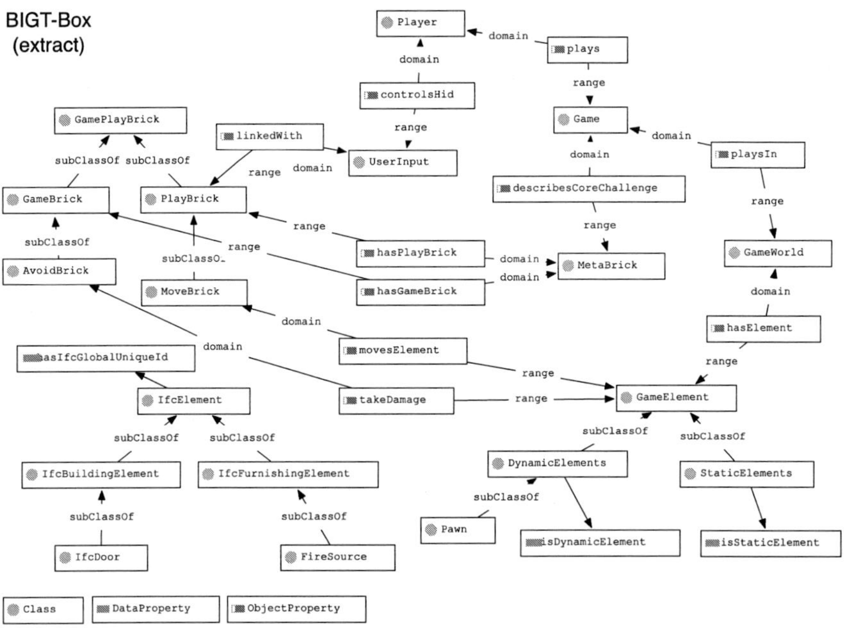

Figure 6-7. Extract of the TBox ontology

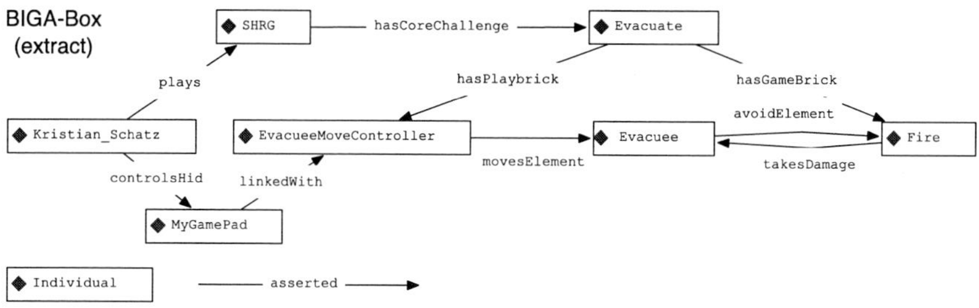

Figure 6-8. Core challenge of the SHRG as ABox

classified with the Data-Property *hasIfcGlobalUniqueId* to store the *GlobalId: IfcGloballyUniqueId* as a String according to the IAI definition (IAI 2007). This *GlobalId* is the connection between conceptual layer and data layer and can be used for deeper analysis to obtain additional using the IFC Java Toolkit.

This approach results in a semantic linking of BIM, 3D modeling and animation and game domain. The use of rules allows deriving dynamic behavior of a building element based on the IFC description. In the case of the presented example the information about the opening operation given by *ifcDoorStyle* is used to add the animation for opening the door to the 3D scene automatically. Beyond that, the ontology includes the concepts and relationships of he "Gameplay Bricks". This approach enable the automation of reasoning procedures and query answering for adding required information about building element specific, interactive behavior and the role within the game.

The description of the core challenge of the SHRG sentence (1):

The player moves his pawn in order to avoid fire and smoke.

can now be formalized with the vocabulary of the developed ontology (Figure 6-8).

From this definition, the inferred model is now able to find answers for most of the "Gameplay Questions". But one aspect is missing, the answer to the question: "What are the elements of the game world?" The *GameElement* individuals can be created and described with their properties. A reasoner can then classify the individuals (red arrows in Figure 6-9) and calculate an inferred model. The *GameElement* individuals should be created based on the *ifcElements* stored in an Ifc file. To add this information it is proposed to use semantic annotation as described in the next section.

Semantic Annotation of Building Elements

Semantic annotation attaches metadata to a document, pointing to concepts and properties in an ontology (Dengel 2012). In the presented approach, semantic annotation is used with goal to enhance the digital building elements that they become "Gameplay Bricks" for an automatic information extraction and instancing to support the game design process directly out of the digital building model.

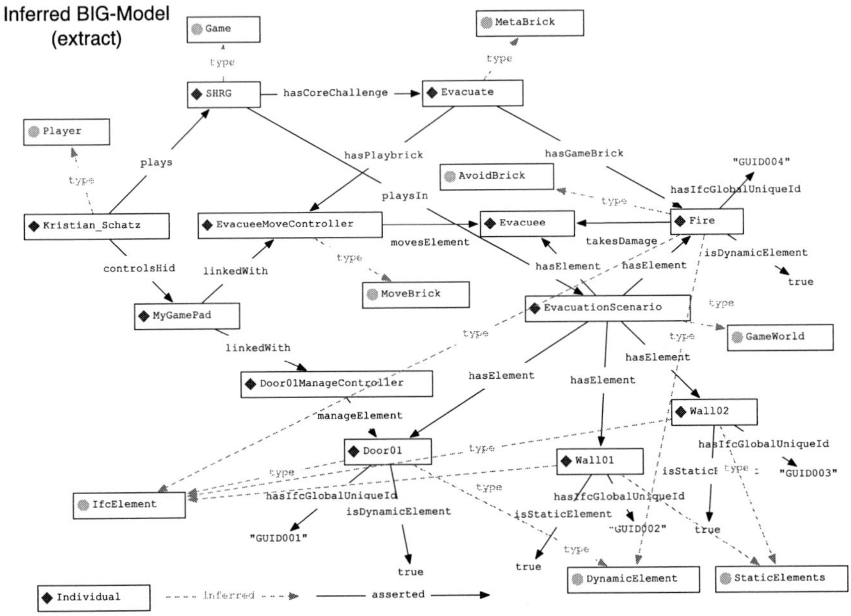

Figure 6-9. Extract of the ABox for the SHRG

In the presented use case, two strategies for semantic annotation are chosen: First, through manual annotation by the user, and secondly by automatic annotation using an inference engine.

For the first strategy, users can create annotations interactively supported by a user interface. The annotations can be added direct in the process of building design with BIM software tools. From this annotated building elements information could be extracted and be added to the knowledge base. Conversely, the existing knowledge can be used to enrich existing building elements by automatic annotation for the second strategy.

Tagging refers to the enrichment of a set of building elements with information about their meaning within game boundaries. For mark-up a vocabulary and format (in the present case the standard format OWL/XML) according to the ontology should be used and restrictions on data types or attributes of classes should be examined.

One of the key issues in the creation of annotations is the question of the storage. Dengel distinguishes three strategies (Dengel 2012):

- *Embedded annotation* (stored directly in the annotated document, read/write access required)
- *Intrinsic annotation* (saved as external resource, enriching the document with reference to the annotations, also write access required)
- *Extrinsic annotation* (stored externally, the annotations refer to the document, no write access to the original document required)

One possibility to use embedded annotation is to store it directly in the IFC document. There should be no confusion with *IfcAnnotation*, which is a graphical representation within the geometric (and spatial) context of a project (IAI 2007). Semantic annotation as proposed could be done with *IfcPropertySingleValue*, which defines a property—single value combination for which the property name, the value (numeric or descriptive) with measure type (and optionally the unit) is given (IAI 2007).

To define a *IfcPropertySingleValue* a Name with type *IfcIdentifier* is required which allows the *IfcPropertySingleValue* to be identified. The Value can be stored as any type of *IfcValue*, e.g. *IfcText* that is an alphanumeric string of characters. With the supported type *IfcText* it is possible to store OWL/XML tags as *IfcPropertySingleValue* embedded in the IFC document.

Source 1 in Figure 6-10 shows the enhanced IFC document for the simple door example. The door is now tagged in way, that a reasoner can classify this element as shown in Figure 6-10.

BUILDING INFORMATION GAMING ENVIRONMENT

Based on the concepts and methods described in the previous sections a Building Information Gaming environment is currently under development. The BIG environment supports the game design process using common file interchange formats and semantic technologies.

During the conversion of the building objects for generating the game objects it automatically takes the semantic annotation of an object into account. The annotation provides information about the type (e.g. static or dynamic) it will be in the game world and how it is connected to gameplay over "Gameplay Bricks" (behavior model). With these features it is possible to add behavior modeling into the Building Information Modeling (BIM) process.

In the simple door example, the building object door must be converted into a dynamic game object because a player-controlled avatar must be able to interact with it in the virtual game world. A door object can have different states (open, closed and unlocked or closed and locked) and behaviors according to these states, which influences the movement possibilities of the avatar.

Components of the Environment

The BIG environment consist of the following components as shown in Figure 6-11:

- *BIM Software*: An Add-in for Autodesk Revit Architecture was developed to provide a graphical user interface (GUI) for the interactive creation of semantic annotations and to communicate with the web services of the BIG.MIR knowledge server. The GUI is explained in detail later on.

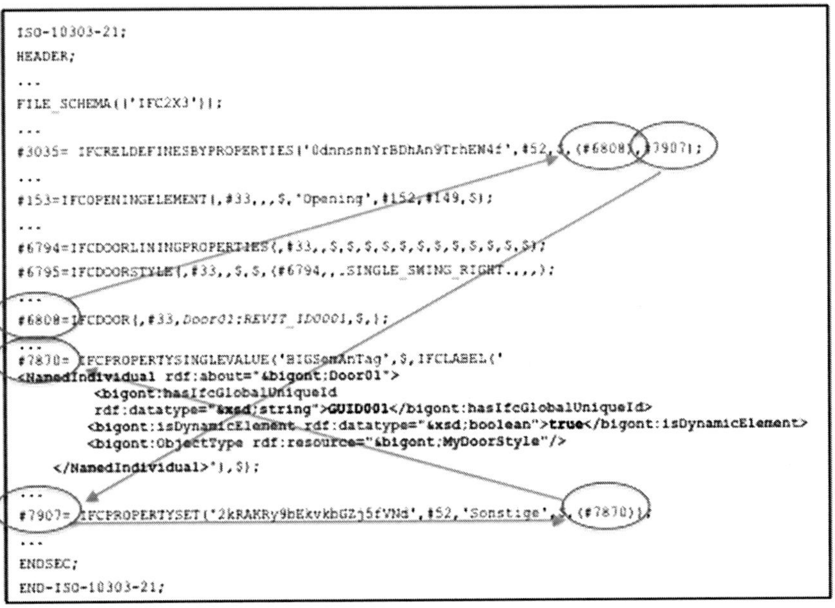

```
ISO-10303-21;
HEADER;
...
FILE_SCHEMA(('IFC2X3'));
...
#3035= IFCRELDEFINESBYPROPERTIES('0dnnsnnYrBDhAn9TrhEN4f',#52,$,(#6808),#7907);
...
#153=IFCOPENINGELEMENT(,#33,,,$,'Opening',#152,#149,$);
...
#6794=IFCDOORLININGPROPERTIES(,#33,,$,$,$,$,$,$,$,$,$,$,$,$);
#6795=IFCDOORSTYLE(,#33,,$,$,(#6794,,.SINGLE_SWING_RIGHT.,,,);
...
#6808=IFCDOOR(,#33,Door01:REVIT_ID0001,$,);
...
#7870= IFCPROPERTYSINGLEVALUE('BIGSemAnTag',$,IFCLABEL('
<NamedIndividual rdf:about="&bigont:Door01">
        <bigont:hasIfcGlobalUniqueId
        rdf:datatype="&xsd;string">GUID001</bigont:hasIfcGlobalUniqueId>
       <bigont:isDynamicElement rdf:datatype="&xsd;boolean">true</bigont:isDynamicElement>
       <bigont:ObjectType rdf:resource="&bigont;MyDoorStyle"/>
    </NamedIndividual>'),$);
...
#7907= IFCPROPERTYSET('2kRAKRy9bEkvkbGZj5fVNd',#52,'Sonstige',$,(#7870));
...
ENDSEC;
END-ISO-10303-21;
```

Source 1. Enhanced *IfcDoor* using semantic annotation

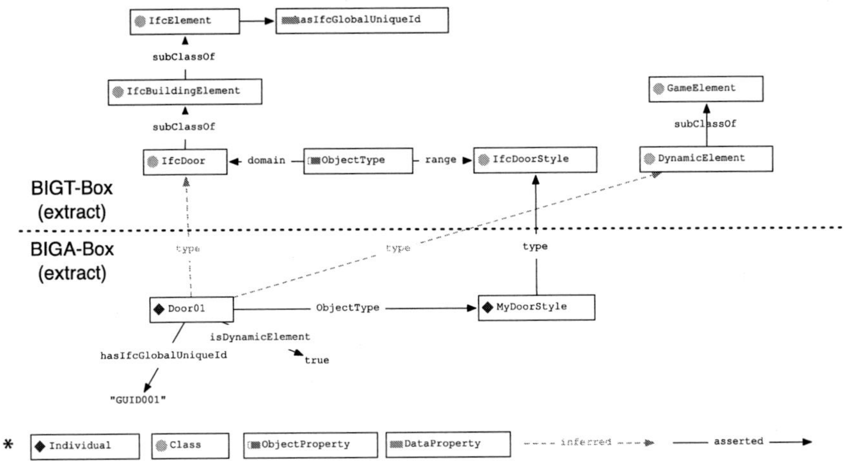

Figure 6-10. Extract of the inferred ontology model based on annotated ifcElement

- *BIG.MIR*: The knowledge server component is realized with the Apache Jena Toolkit. It provides a web service interface to upload the TBox ontology model as OWL file e.g. developed with Protégé-OWL as well as the IFC and FBX data and to map the received data to the ontology model.

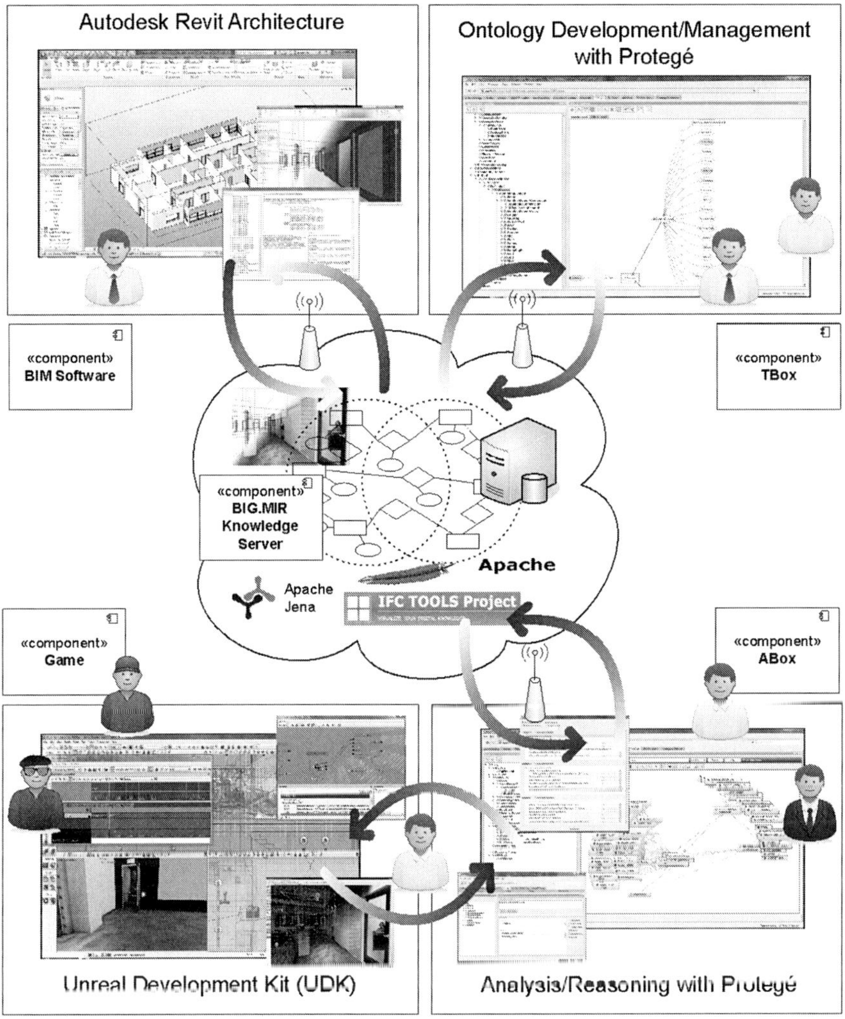

Figure 6-11. Building Information Gaming environment

It includes features to query the ontology as well as to access data stored in the IFC file over the Ifc Java Toolkit (Ifc Tools Project 2013) and in the FBX file over the FBX SDK (Autodesk 2012). The resulting ontology model can be downloaded to the client as OWL file for further analysis, reasoning or query answering with Protégé-OWL.

- *BIG.ONT*: The knowledge base. Covers the conceptual view of the integrated domains (TBox) and the asserted model (ABox) as OWL files for further processing with Protégé-OWL or BIG.MIR.

- *Game Development Kit*: Modeling the game world based on the game design documents provided as OntoGraf diagrams and the building objects, which are now available as game content via the FBX exchange format.

Graphical User Interface

The elements of the GUI are organized in two main sections (see Figure 6-12):

- *BIG.BASE*: Provides functionality for uploading and downloading of the ontology model to and from the BIG.MIR webservice.
- *BIG.DESIGN*: Opens a GUI for graph-based, semantic annotation of the digital building data according to (Schatz and Rüppel 2014).

A TBox ontology model, which is submitted to the BIG.MIR knowledge server, could be selected to provide the vocabulary for the current project. With this vocabulary the engineer is now able to formulate the core challenge of the game and to create the individuals of the digital building model and their relationships based on the IFC structure graph (see Figure 6-13). The functions shown in Figure 6-13 will now be explained:

1. Opens a window for graph-based, semantic annotation.
2. Functionalities for creating and reusing individuals based on the contextual ontology model.
3. Functionalities for creating relationships between individuals (object properties) of the contextual ontology model close to natural language (subject-predicate-object pattern).
4. Functionalities for creating relationships between individuals and literals (data properties).

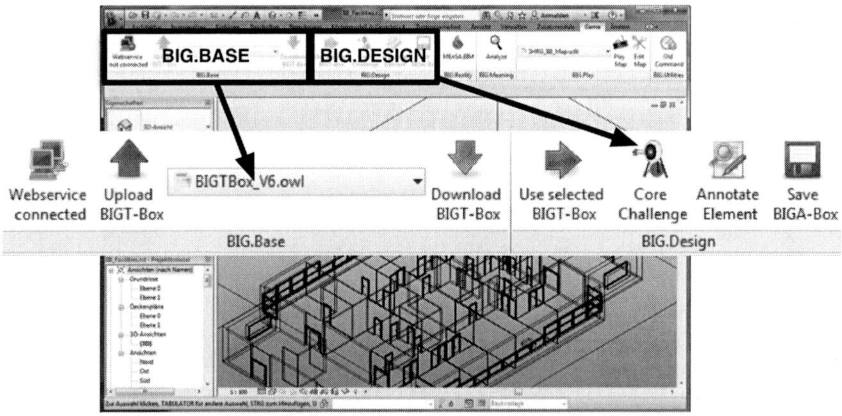

Figure 6-12. Revit Add-in Ribbon Panel

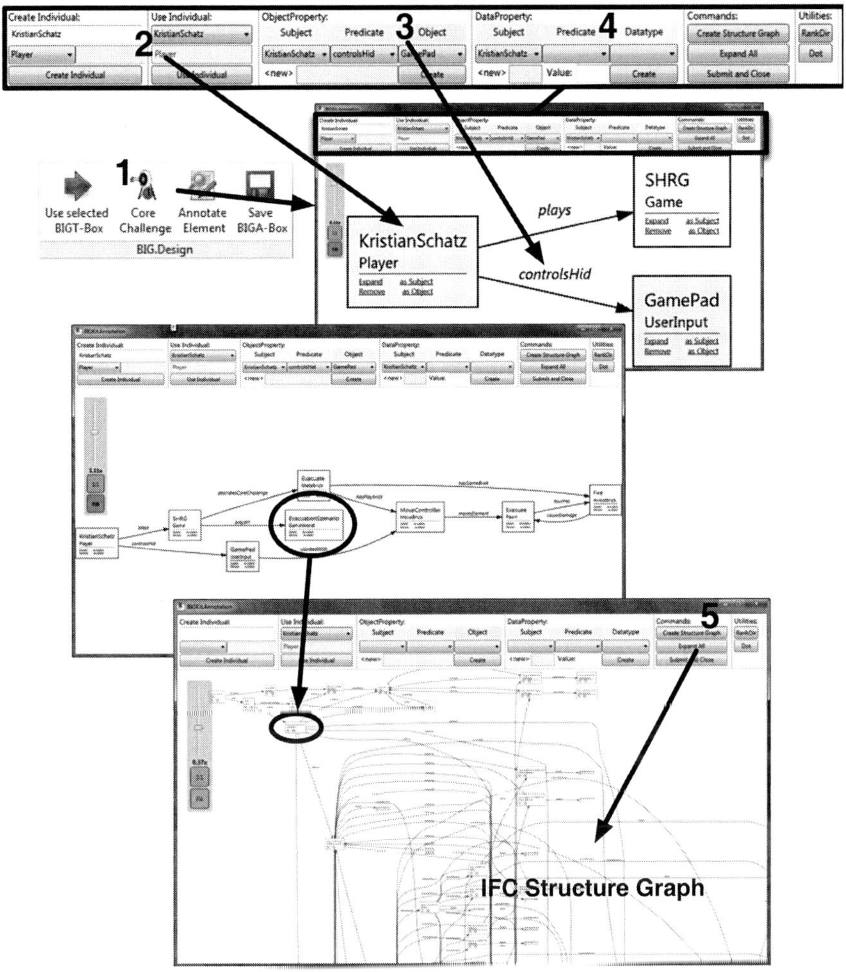

Figure 6 13. Semantic modelling of the core challenge

5. Functionalities for creating a semantic network out of digital building model based on the IFC relationships.

After the core challenge is modelled and the structure graph is generated the engineer can know go more into detail and enhance the existing digital building objects by graph-based, semantic annotation.

The annotation process is demonstrated in Figure 6-14:

(1) Opens a window for graph-based, semantic annotation with focus on the selected building object in the 3D view of Revit.

(2) Functionalities for creating and reusing individuals based on the contextual ontology model are used to model the game object definition.

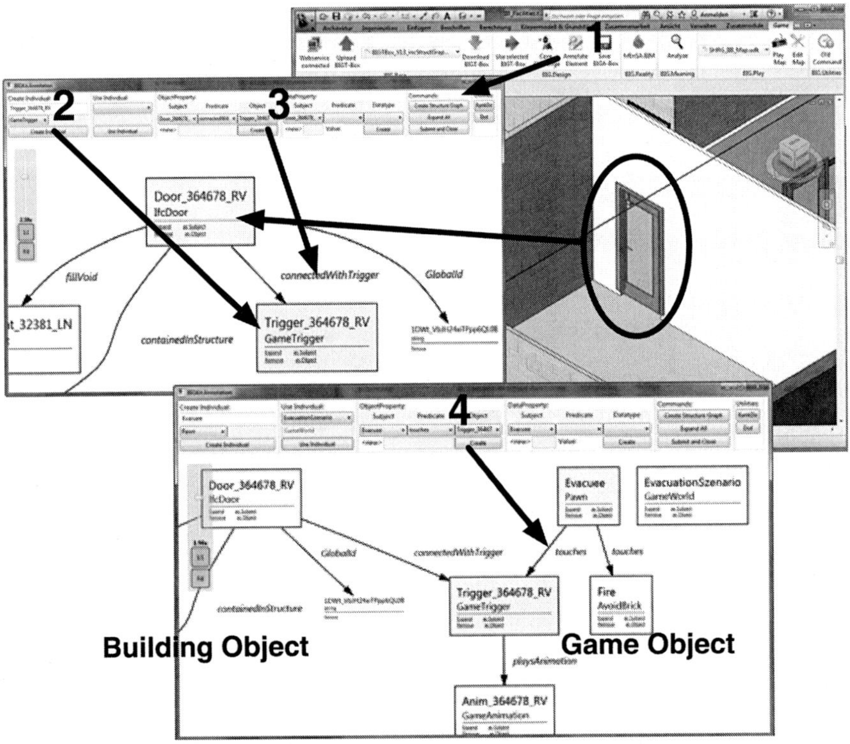

Figure 6-14. Semantic annotation of the building object

(3)+(4) Functionalities for creating relationships between individuals are used to connect the game object definition with the digital building object for behaviour modelling in the game design context.

SERIOUS HUMAN RESCUE GAME DESIGN MODEL

The BIG environment was used to support the design process of the SHRG. The base for the game scenario was a digital building model designed with Autodesk Revit. By graph-based, semantic annotation BIM objects could be extended to aspects of meaning in the context of the game (behaviour modelling). The annotated building data could be further processed with the BIG.MIR web service and analysed with Protégé-OWL.

A time-and cost-consuming generation of additional game design documents (e.g. UML or entity relationship diagrams) could be avoided, as with OntoGraf, a functionality of Protégé-OWL, the essential aspects of computer game model could be represented visually as semantic network (see Figures 6-15–6-17). A student in the role of a game designer was able to model the game scenario based on the information with the unreal development kit.

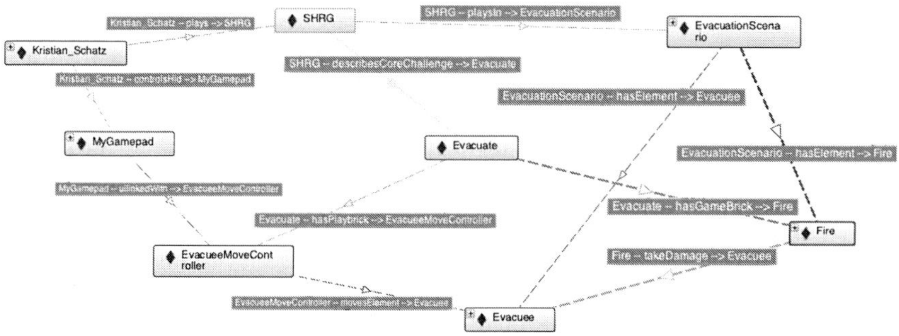

Figure 6-15. ABox: CoreChallenge (OntoGraf)

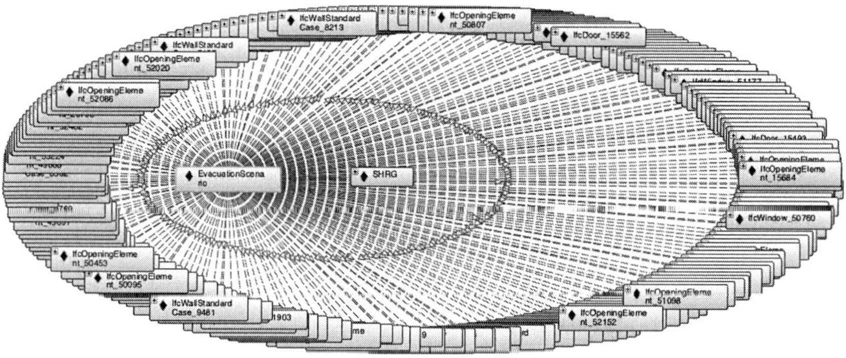

Figure 6-16. ABox: SHRG scenario individuals (OntoGraf)

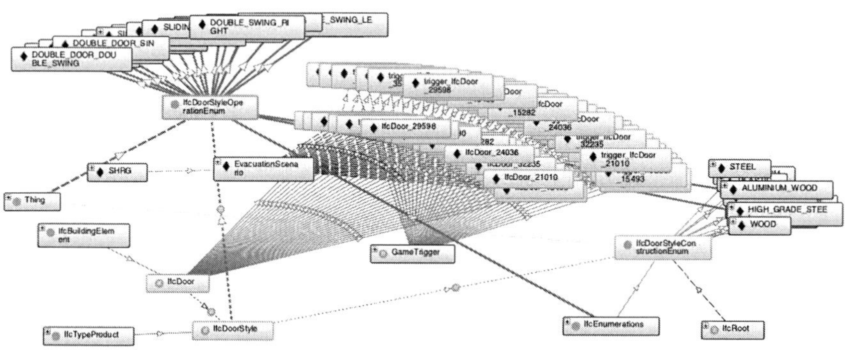

Figure 6-17. ABox: Aspects of the enhanced IfcDoor objects (OntoGraf)

CONCLUSIONS AND OUTLOOK

This chapter introduces an ontology-based approach for integrating building information modeling into serious gaming environments. The challenge was to displace the interpretation of information from the developer into a computer system. To solve this problem an ontology-based approach was chosen to automate the game content creation as shown in Figure 6-18.

The developed BIG ontology is used to integrate data stored in different data sources for BIM and 3D content creation and to enhance this data with information about gameplay behavior. The ontology acts as a conceptual layer that provides reasoning and query answering over the integrated data model. This feature supports the generation of game content based on the digital building model in phase 2 of the BIG design process. This helps to save time and expenses for developing Serious Games in the field of AEC.

One advantage of the presented approach is, that to make use of the developed methods and concepts no changes in the IFC schema is required. Every *IFCElement* with a *GlobalID* could be a resource (subject/object) including their relationships (properties) in the semantic network and could be enhanced with additional meta-data based on the vocabulary provided by an ontology model of the domain of interest. The presented use-case shows, that for behavior modeling in the context of the SHRG design process this approach could be used already successfully.

It is assumed that this combination has the potential to set a technology trend in using BIG for all Serious Games with a relation to the real built world. BIM software tools can provide the bridging technology between the real world and the game scenario.

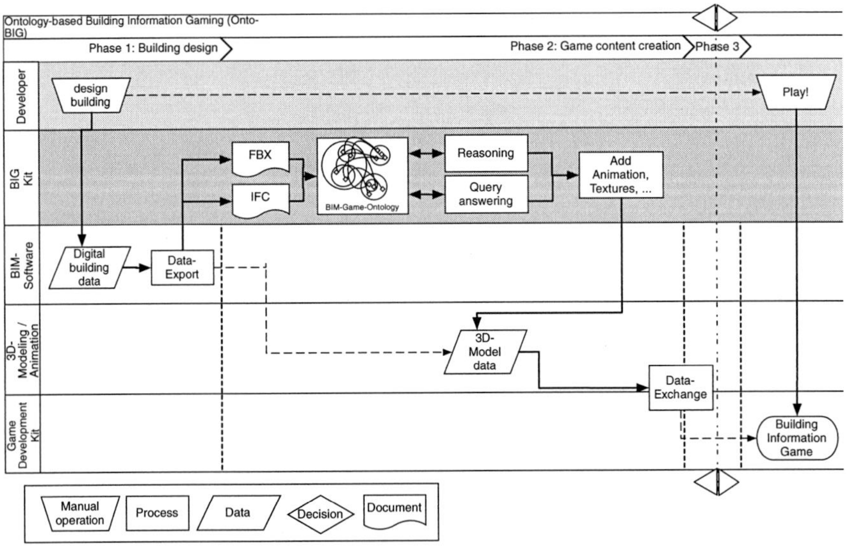

Figure 6-18. Semantic technologies for Building Information Game design

References

Autodesk. (2012). *Autodesk FBX 2012 SDK Documentation*, Available online: http://download.autodesk.com/global/docs/fbxsdk2012/en_us/index.html, Last accessed: March, 2012.

Calvanese, D. et al. (2011). *The MASTRO system for ontology-based data access*, Semantic Web Journal, 2(1):43–53, 2011, available online: ~http://www.inf.unibz.it/~calvanese/papers/calv-etal-SWJ-2011.pdf, Last accessed: March, 2012.

Claypool, K. and Claypool, M. (2007). *On frame rate and player performance in first person shooter games*, Journal Multimedia Systems, Springer Berlin / Heidelberg ISSN: 0942-4962.

Dengel, A. (Hrsg.) (2012). *Semantische Technologien: Grundlagen Konzepte, Anwendungen*, Springer Heidelberg 2012, ISBN 978-3-8274-2663-5.

Djaouti, D., Alvarez, J., Jessel1, J.-P., and Methel, G. (2007). *Play, Game, World: Anatomy of a Videogame*, in proceedings of CGames'07 conference, France, available online. http://www.gameclassification.com/files/articles/[cgames07]_videogames_anatomy.pdf. Last accessed: May, 2013.

GOP. (2005). *Game Ontology Project (GOP) Website* http://www.gameontology.com/index.php/Main_Page, Last accessed: May, 2013.

Gregory, J. (2009). *Game Engine Architecture*. Wellesley: A K Peters ISBN 978-1-56881-413-1.

Harteveld, C. (2011). *Triadic Game Design—Balancing Reality, Meaning and Play*, Springer London, February 2011, ISBN 978-1-84996-157-8.

Hitzler, P., Krotzsch, M., Rudolph, S., and Sure, Y. (2008). *Semantic Web—Grundlagen*, Springer Heidelberg, 2008, ISBN 978-3-540-33993-9.

Horridge, J. (2011). *A Practical Guide To Building OWL Ontologies Using Protégé 4 and CO-ODE Tools Edition 1.3*, Available online: http://owl.cs.manchester.ac.uk/tutorials/protegeowltutorial/resources/ProtegeOWLTutorialP4_v1_3.pdf Last accessed: March, 2012.

Huizinga, M. (1938). *Homo Ludens – Vom Ursprung der Kultur im Spiel*, 22. Edition, Rowohlt, Hamburg, April, 2011, ISBN 978-3-499-55435-3.

IAI. (2007). *IFC2x Edition 3 Technical Corrigendum 1* Online Documentation http://www.buildingsmart-tech.org/ifc/IFC2x3/TC1/html/, last accessed May 2013.

IFC Tools Project. (2013). Ifc Java Toolbox—Online Documentation http://www.ifctool-sproject.com, last accessed May 2013.

McGonigal, J. (2012). *Reality Is Broken—Why Games Make Us Better and How They Can Change the World*, The Penguin Press, New York ISBN 978 1 101 17549 2.

Noy, N. F. and McGuinness, D. L. (2000). *Ontology Development 101: A Guide to Creating Your First Ontology*, Available online: http://protege.stanford.edu/publications/ontology_development/ontology101.pdf. Last accessed: May, 2013.

Prensky, M. R. (2011). *From Digital Natives to Digital Wisdom: Hopeful Essays for 21st Century Learning*, Corwin, January 2012, ISBN 978-1452230092.

Rüppel, U. and Schatz, K. (2011). *Scenario Builder for generating realistic serious game scenarios for fire safety applications*, *Proceedings of the EG-ICE Workshop 2011*, 6–8, July 2011, Twente University.

Rüppel, U. and Schatz, K. (2011). *Designing a BIM-based serious game for fire safety evacuation simulations*, Advanced Engineering Informatics, Volume 25, Issue 4, October 2011, Pages 600–611, ISSN 1474-0346.

Schatz, K. and Rüppel, U. (2012). *Ontology-based approach for integrating building information modeling into serious gaming environments*, In: *Proceedings of the 14th International Conference on Computing in Civil and Building Engineering*, ISBN 978-5-93093-877-1, Moscow, June 2012.

Schatz, K. and Rüppel, U. (2014). *Sticky "nodes" for BIM: Graph-based, semantic annotation of digital building data*, Accepted abstract. Will be published in: *Proceedings of the 21st International Workshop: Intelligent Computing in Engineering 2014 July 16-18*, 2014 in Cardiff, The United Kingdom.

SruckenSchmidt, H. (2011). *Ontologien Konzepte, Technologien, und Anwendungen*, Springer-Verlag Berlin Heidelberg 2011, ISBN 978-3-642-05403-7.

UDK. (2013). *Epic Games Unreal Development Kit (UDK)* Online Documentation http://www.unrealengine.com/en/udk/documentation/, last accessed January 2013.

W3C. (2009). *OWL 2 Web Ontology Language Document Overview, W3C Recommendation* 27 October 2009, available online: http://www.w3.org/TR/owl2-overview/, Last accessed: March, 2012.

CHAPTER 7

Fieldwork-Based Method for End-User Engagement in Domain Ontology Development

L. L. Olde Scholtenhuis*
T. Hartmann†

Abstract: *This chapter introduces a fieldwork-based method for end-user engaged ontology development. Our approach complements existing methodologies that, despite their aims to consult end-users and domain experts, have not yet formally spelled out steps for end-user engagement in detail. The method we introduce is based on ethnographic-action research. This grassroots approach allows developers to create feasible ontologies that closely fit end-user needs. Being based on fieldwork, this method extends conventional desk-research and expert-panel methods for data acquisition, conceptualization and validation to more meaningfully derive objects and relations that closely match the work routines of practitioners that eventually use ontologies. This chapter reviews existing post-development methods for end-user involvement and shows how these methods can be complemented with a grounded approach such as ethnographic-action research. We demonstrate the method by developing a semi-formal domain ontology for the subsurface utility construction domain.*

INTRODUCTION

Ontologies form the knowledge base of information systems and are defined as formal and explicit specifications of shared conceptualizations (Studer, Benjamins

*PhD-student, Department of Construction Management and Engineering, University of Twente, P.O. Box 217, Enschede, The Netherlands; email: l.l.oldescholtenhuis@utwente.nl

†Associate professor, Department of Construction Management and Engineering, University of Twente, P.O. Box 217, Enschede, The Netherlands; email: t.hartmann@utwente.nl

et al. 1998). *Conceptualization* refers to abstracted and simplified views (Guarino, Oberle et al. 2009) on the reality, or more carefully formulated, the universe of discourse (Turk 2001) we wish to model (the latter term does not imply a real or tangible world, but merely refers to a perceived reality). *Shared* refers to the multiple views that an ontology should be able to represent. This multi-view representation is necessary, since ontologies that only represent an individual viewpoint are too narrow, being of little relevance to the broader public that eventually uses the ontology. In contrast, too broad and generic ontologies representing concepts beyond a specific domain of discourse might contain redundancies. Redundant concepts unnecessarily complicate an ontology's structure, create ambiguities and make the ontology more difficult to understand by its end-users. Furthermore, ontologies are *formal* and *explicit*. These terms refer to the aim of ontologies to describe concepts clearly in a computer interpretable format. More pragmatically, this means that ontologies are comprised of objects with time-bounded relations, properties and attributes. Rules and axioms are used to relate various objects that, for example, compose events, instants, processes and states (Chandrasekaran, Josephson et al. 1999).

The scope and purpose of an ontology influence how developers conceptualize various objects, relations and properties. Consequently, ontology developers can represent a single concept in alternative ways. In practice, scopes vary from the general to specific. Most abstract and generic ontologies have a broad scope and are usually called upper-level or top-level ontologies. They have a holistic view on a universe of discourse and use just a few generic concepts to describe this. For some purposes, top-level concepts are too imprecise. Practitioners that, for example, use ontologies in information systems require a more specialized and detailed conceptualization. For this reason, developers often create bounded (Kishore et al. 2004) and more concrete conceptualizations; so called domain ontologies.

This chapter focuses on discussing methods for developing domain ontologies. Such ontologies ideally allow information systems to think and communicate in the language of end-users such as engineers (Katranuschkov et al. 2003). From an interpretive stance, such domain end-users observe, interpret, label and give meaning to phenomena in distinctive ways. So, what is highlighted by one individual might be irrelevant for another. Take, for example, the concept *bridge*. This seemingly straightforward object can be looked at in different ways. In the context of civil engineering, for example, architects look at this concept using lower-level concepts such as material, colors and shapes, while structural engineers could perceive a bridge in terms of load bearing capacity, materials and construction methods. On the other hand, traffic engineers might highlight concepts such as driving lanes, traffic flows and average driving speed.

The various perceptions of the individuals are incomplete and bounded by rationality (Simon 1976). Due to the bounded cognitive capabilities and varying perspectives on reality, it would be impossible for a modeler to create a conclusive model that perfectly represents all possible viewpoints on a domain. In any model, abstracted concepts hence highlight features that are important for one, but ignore details and complexities for other purposes. At best ontological models therefore only represent a partial understanding of the real world (Turk 2001). What follows from this is that perspectives of developers influence the eventual shape of an ontology as they decide which parts of a domain of discourse they highlight or exclude (Björk 1992). Therefore, developers need to take due care in investigating the multiple end-user viewpoints on a domain.

Many existing approaches to ontology development acknowledge the need to sufficiently study end-user perspectives. The methodologies, for example, rely on domain expert interviews, panels, brainstorm sessions and end-user boards. One of the problems of these methods is that end-users often have difficulties in verbalizing domain knowledge themselves due to their limited conceptualizing experience. Secondly, domain experts seem distanced from end-user work practices and routines, making it hard to meaningfully conceptualize important domain knowledge. As a result, it becomes difficult to prevent developers from creating comprehensive ontologies that contain more knowledge than needed or miss details for application in a specific domain. To address this problem, we introduce fieldwork-based end-user engagement. Our method provides formal steps for examining and conceptualizing domains through close interaction with the domain.

This chapter is structured as follows: We first summarize how six existing ontology development methodologies formalize end-user engagement for common development steps such as data acquisition, conceptualization and validation. Next, we argue that existing top-down and post development validation methods can be strengthened by using fieldwork. Subsequently, we introduce a formal method—based on qualitative fieldwork techniques such as ethnography and action research—that enables developers to better engage with end-users. Finally, we exemplify how we used this to develop a domain ontology for the urban subsurface reconstruction project domain.

END-USER ENGAGEMENT IN ONTOLOGY DEVELOPMENT

Ontology developers create models that represent general concepts such as actions, time, physical objects and beliefs (Russell et al. 1995). Steps for creating such abstractions include scoping, data acquisition, concept exploration, definition of terms, structuring and validation. To effectively execute these steps, the process includes the roles of ontology developer, domain expert and end-user. Ontology developers are usually experienced in building models in computer interpretable languages. Further, domain experts are knowledgeable about a domain of discourse. Finally, end-users are practitioners that use information systems of which ontologies are part. Unfortunately, many expert and end-user involvement methods seem to inadequately deliver valuable input in the ontology development process. This section elaborates this point by providing an overview of end-user involvement in the methodologies TOVE, Enterprise, Methontology,

DILIGENT, Ontology Engineering Methodology and the handbook-based ontology development methodology. We summarized our findings in Table 7-1. We refer to Gómez-Pérez (2004) for more comprehensive comparison of existing methodologies.

In the Toronto Virtual Enterprise (Gruninger and Fox 1995) methodology (TOVE), industry partners provide input for the scope of the ontology. To this end, developers invite end-users to formulate motivating scenarios. To incorporate these in the ontology design, developers create informal competency questions. Answers to these questions define the knowledge that the ontology should be able to provide. Based on the questions, developers identify concepts, requirements, terminologies, definitions and constraints. To validate, developers themselves test whether the ontology, written in computer interpretable language, is technically complete. Additionally, they check whether the ontology answers the competency questions. In sum, TOVE includes end-user's competency scenarios to develop their ontology. However, TOVE misses detailed descriptions of how end-users can convert their knowledge into formulating scenarios effectively.

A methodology that focuses on the creation of an ontology in natural language is Enterprise (Uschold and Gruninger 1996). The approach suggests to first define a scope. Subsequently, it focuses on producing definitions through a middle-out approach. This means that developers move back and forth between identification of higher and lower-level concepts. In the steps that follow, developers review the concepts and devise a meta-ontology. For concept exploration, Enterprise proposes to organize brainstorm sessions with domain experts. Unfortunately, Ushold and Gruninger (1996) do not provide great detail about how end-users and developers could conceptualize in brainstorm sessions. Further, as the developer mainly relies on expert knowledge, Enterprise limits the developer in validating whether conceptualizations are also shared by end-users.

Furthermore, a methodology originating from the domain of chemistry is Methontology (Gómez-Pérez, Fernández et al. 1996; Fernández-López 1999). Methontology proposes to capture knowledge and specify requirements in tree-structures and tables (called intermediate representations). As a next step, domain experts are asked to verify these representations before they are implemented in computer interpretable language. In the last step, the methodology proposes to evaluate the ontology using formal consistency checks and validation interviews. Unfortunately, Methontology does not provide formal steps describing expert's verification sessions and evaluation interviews. Further, it does not elaborate how developers could check whether the expert viewpoints actually represent the perspectives of the ontology end-users.

Another method for end-user engagement is described in DILIGENT (Pinto, Tempich et al. 2009). The DILIGENT methodology is based on the assumption that ontologies could be developed in a decentralized way—i.e. without developers gathering physically. DILIGENT was developed to address the lack of non-expert software designers in the process of ontology design and allows users to adapt ontologies to their local needs. In the first step of the process that DILIGENT

Table 7-1. Comparing existing methodologies for domain ontology development

Methodology	*TOVE*	*Enterprise*	*Methontology*	*DILIGENT*	*Ontology Engineering Methodology*	*'Handbook-based ontology'*
Reference	*(Gruninger and Fox 1995)*	*(Uschold 1996; Uschold and Gruninger 1996)*	*(Gómez-Pérez, Fernández et al. 1996; López, Gómez-Pérez et al. 1999)*	*(Pinto, Tempich et al. 2009)*	*(Sure, Staab et al. 2009)*	*(Hsieh et al. 2011)*
Methods for knowledge acquisition and concept exploration	Competency questions	Brainstorming	Expert meetings, interviews, document analysis	Domain experts involvement, end-users can propose amendments	Ontology re-use, automated document analysis, domain expert consultation	Semi-automatic domain handbook analysis
Validation methods	Formal competency questions	Check model against purpose and user requirements	Consistency and redundancy check, expert inspection	Local adaptations, board reviews	Technical evaluation on consistency, user-focused evaluation	Expert workshops
Expert and end-user involvement	Industry partners formulate 'motivating scenarios'	Domain experts participate in brainstorm	Domain experts verify representational drawings and participate in post-development interviews	Experts, developers, users create initial ontology, end-users and boards suggest and accept adaptations	Experts consultation and interviews, users involved in ex-post evaluation	Domain experts are only involved in upper level concepts definition and review and evaluation

prescribes, a small team of domain experts, end-users and software developers create an initial version of a shared ontology. Then, the team publishes the first shared ontology and allows end-users to adapt the version to their local needs.

Subsequently, a board of developers evaluates local changes and decides whether they also incorporate these in a next version of the shared ontology. The local adaptation and board update can be seen as validation steps that focus on end-user satisfaction and a technical consistency. The steps in the DILIGENT-process are iterative: developers go back and forth between development, evaluation and local adaptation. Although end-users can have great input, DILIGENT does not provide detailed steps supporting conceptualizing efforts of developers and end-users.

The next approach is Ontology Engineering Methodology (Sure et al. 2009). This methodology describes how to develop and maintain ontology-based knowledge management applications. It integrates the development of an ontology and the information system in which it will be used. The methodology has five general steps. First, developers set requirements for the ontology (feasibility study). After that they describe a first semi-formal version of an ontology (kickoff) and translate this version in a formal target ontology (refinement). What follows are evaluation, application and evolution steps. For concept exploration, Ontology Engineering Methodology suggests to re-use existing ontologies, conduct automated document analysis or conduct expert interviews. In the validation step, Sure et al. (2009) propose to conduct technology-focused and end-user-focused evaluations. The technology-focused evaluation directs at language conformity, consistency and proper use of formal ontological language, while end-user-focused evaluation aims to validate the ontology from the viewpoint of an end-user. Despite referencing to these end-user engaged methods, this methodology provides little information about how the steps should actually be conducted.

Finally, Hsien et al. (2011) introduce a way to develop an ontology semi-automatically. To this end, they used information systems to extract domain concepts from engineering handbooks. The primary focus of this methodology is to develop a draft earthquake engineering domain ontology. Simultaneously, Hsien et al. (2011) aim to limit time-consuming participation of domain experts. Their methodology therefore uses computers to semi-automatically study engineering handbooks and to develop a glossary of terms. Developers are then proposed to analyze tables of content, definitions and book-indices. From these documents, they extract and organize various concepts and instances. As this methodology's goal is to reduce expert involvement, it implicitly ignores the important task of end-user engagement. The authors admit this by concluding that domain experts were still necessary for post development activities such as revising, weighing and filtering concepts.

The previous ontology development methodologies show various approaches to end-user engagement ranging from motivating scenarios, to re-use of ontologies, brainstorm sessions, interviews and board updates. Although these methods provide relevant input for ontology development and validation, they risk creating inadequate abstractions of end-user domain knowledge. One reason for this might be that existing end-user engagement methods are spelled out only limitedly. Further, it seems that reliance on expert methods does not bring developers close enough to the intricacies within the domain of interest. As experts are often distanced from end-user work tasks and routines, their knowledge seems only partially adequate to create comprehensive domain ontologies. Finally, also end-users have difficulties contributing domain knowledge since these practitioners often have limited experience in verbalizing their own knowledge. Resulting ontologies may contain specification failures, poor conceptualization and consensus problems. Improving representativeness of these ontologies eventually involves consuming validation and rework such as removing redundancies and identifying initially overlooked main concepts.

In reaction to the outlined need to improve representativeness of current ontology development methodologies, this chapter calls for more effective end-user engagement. Therefore, developers need to investigate domains more closely to abstract end-user knowledge. This could be done by applying bottom up strategies (El-Diraby 2012). One such strategy involves fieldwork. When conducting fieldwork, developers step away from their desk to spend a significant amount of time in the domain of study. Here, they to learn about practitioners' working culture, practices and routines and find out about domain intricacies. Developers' presence in the field can assist practitioners in verbalizing their domain knowledge, while they can simultaneously make abstractions based on their own real-life observations.

Another benefit that follows from direct interaction between ontology developers and the domain is that it offers a possibility to implement and validate the ontology (including its information system) directly within the domain: when developers observe this first ontology implementation, they can immediately identify minor changes that are necessary to increase its user fitness. In contrast with desk-research development, such necessary changes can be identified and addressed quickly whilst ontology developers are actually in the field themselves.

In sum, we argue that fieldwork provides an opportunity to better ground ontological concepts and relations in various end-user perspectives. In our view, this significantly increases user-fitness and representativeness and reduces the likelihood that too-large or irrelevant ontologies are developed. To explain how fieldwork can actually be used in ontology development, the next section spells out our formal steps for a fieldwork-based method.

FIELDWORK-BASED ONTOLOGY DEVELOPMENT METHODOLOGY

Developers can explore domain practitioners' viewpoints by deploying a fieldwork-based development approach such as ethnographic-action research (c.s. Hartmann, Fischer et al. 2009). This methodology allows ontology developers to adequately conceptualize from an end-user perspective. To this end, it combines explorative and grounded ethnographic research with more engaged action research. We sequentially discuss both research types and integrate these in a formal model for end-user engagement.

Ethnography (Phelps and Horman 2010) originates from cultural studies that aim to grasp socially complex phenomena. Researchers conducting ethnographic studies try to become insiders within a culture as they try to investigate a phenomenon of interest from a close distance during a long-term period of fieldwork. Ethnographers usually join a group, organization or community—e.g. a contractor, project management team, or engineers—to become part of practice and to 'go native'. To discover the concepts and perspectives, ethnographers enter a domain and learn something, try to make sense out it, verify whether the interpretation made sense in its context, refine the interpretation, and so on (Agar 1996, pp. 62). In contrast with existing desk-research that does not capture detailed end-user knowledge, ethnography allows for exploration of phenomena that are hard to understand for domain outsiders. In essence, ethnography allows developers to acquire a holistic, end-user-oriented view on a domain, its concepts and language.

Additionally, action research focuses on studying changing practices. In action research, researchers actively participate in a social environment in which they introduce a planned change. Action researchers could, for example, study how an implemented new technology changes work practices by observing a work practice before, during and after the implementation of the technology (Hartmann, Fischer et al. 2009). The researchers investigate these social and technological phenomena by working together with the population of interest, forming "co-participants of enquiry of a change process" (Baskerville 1999). As ontological models are implemented within information systems, developers can use action research to validate an ontology in its domain. Developers can do this by observing how end-users use the information system and, based on this, identify whether additional concepts and relations need to be incorporated in their ontology.

Ethnography and action research complement one another: while the former allows developers to better understand the domain that they model, the latter allows them to implement and evaluate the model in a real-life setting. This subsequently helps developers to adapt ontologies to the needs of practitioners. Hartmann et al. (2009) integrated both approaches and created the ethnographic-action research cycle. They demonstrated it for development of construction management supporting information systems.

We argue that ontology developers can use ethnographic-action research to enhance the feasibility and user-fitness of ontologies. The methodology amends existing desk-research and expert approaches as it provides a more holistic method for domain exploration. As suggested previously, this fieldwork-based approach brings developers closer to the end-user perspective. This subsequently allows them to support practitioners in verbalizing knowledge and to ground conceptualizations in actual observations of practice. Additionally, it offers a way to iteratively validate and improve the domain ontology. Through this, intricate ontology details can be identified while developers can also find out which insignificant generalities to discard.

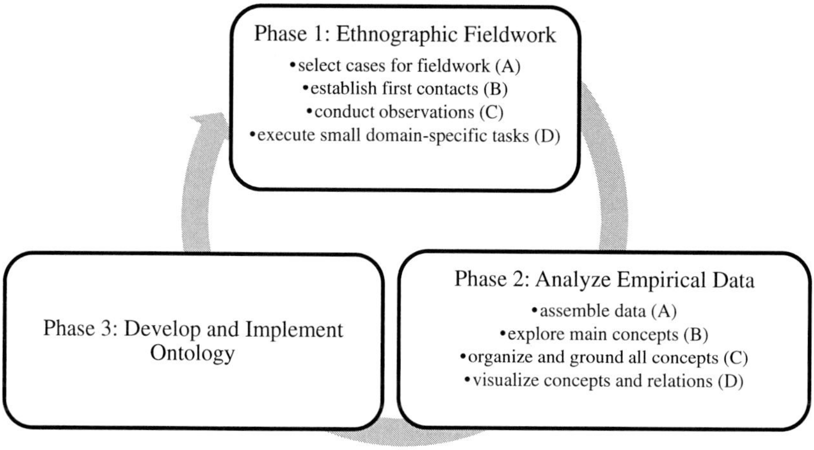

Figure 7-1. Ontology development process, adapted from Hartmann et al. (2009)

We adapted the model from Hartmann et al. (2009) for the context of ontology development in Figure 7-1. In the first step of our adapted ethnographic-action research cycle, developers physically enter a field of study to identify practitioner's routines and work practices. They try to learn the local language and then use this language to describe instances, terms and concepts that are part of practitioners' domain perceptions. After the developers arranged these objects, they arrive at a first ontology version, containing various instances, concepts and categories that describe the domain from multiple viewpoints. In subsequent steps, developers can evaluate their ontology by implementing it in a prototypical information system. In the steps that follow, developers use action research to observe how end-users use the prototype in practice. These observations then help to amend or refine concepts represented in the ontology. In the next paragraphs, we elaborate these ontology development steps in detail.

Phase 1: Ethnographic Fieldwork

In the first step of our method developers select their population of interest based on the scope and purpose of the ontology. Then they establish first contacts and make practical arrangements in the field. We elaborate on both steps below.

Select Cases for Fieldwork

An in-depth domain study requires developers to work closely together with practitioners. To guide the identification of a project case that meaningfully contributes to the ontology, our method suggests using the ontology's scope and purpose. Additionally, we propose to search for cases representing every day and commonplace situations of a domain. Conceptualizations from this case are assumed to be informative about the domain as a whole (Yin 2003). As scopes and purposes can vary amongst domain ontologies, this also implicates that different project cases can be studied. For example, narrow scoped ontologies require a selection of cases in a specialized field. Such fields can be unique construction projects, a specific civil engineering discipline, or a type of construction organization. On the other hand, conceptualizing more general domains like 'the civil engineering field' or 'general construction projects,' require a set of participants that have a wider range of expertise.

Establish First Contacts

After identifying case projects, we suggest to establish first contacts with the practitioners that eventually use the ontology. This step aims to align expectations between practitioners and the ontology developer. Here, developers explain the purpose of their fieldwork to gain access to the domain of the end-user. We also suggest developers to make practical appointments about the time frame of their study, the type of phenomena they wish to observe and the expected deliverable.

As developers established contact with practice, they can now start collecting data. For example, developers could ask whether they can observe meetings, conduct interviews and collect documentation. At the start, the fieldwork is non-intrusive: the developer does not yet participate in practice. At later stages, however, the developer increasingly takes an active position and participates in daily tasks and routines.

Conduct Observations

In this step, developers explore practice and try to make sense of practitioners' domains of discourse. To explore the field in a non-intrusive way, we suggest using various methods for observation. Developers could, for example, visit construction sites, stakeholder meetings, or gatherings at offices. At these places, we advise to conduct participant observations (Jorgensen 1989), take out ethnographic interviews (Spradley 1979), study project documentation and have informal chats with practitioners. It is important that developers register their first observations in the greatest detail possible. Possible ways to do this are by making field notes, recording meetings on audio or video, taking pictures and collecting meeting minutes. Once the fieldwork started, developers should keep track of their conceptualization ideas. They can do this by writing down unfamiliar words, concepts, instances and terms in a fieldwork-diary. This diary can later guide analysis tasks.

Execute Small Domain Specific Tasks

After the initial exploration of the domain, the next step of the methodology proposes developers to engage with practice more actively. To this end, developers can execute various small engineering or management tasks. This brings developers closer to the insider position and allows them to better reflect on the domain. However, as this active fieldwork is time consuming, developers need to carefully think of how a practical task contributes to the eventual ontology. We therefore suggest developers to continually consider how fieldwork participation fits within the scope of the ontology.

Phase 2: Analyze Empirical Data

In this phase, we propose developers to make their first efforts to conceptualize a domain by abstracting and organizing concepts for qualitative field data. This analysis phase should run almost in parallel with fieldwork activities (phase 1) and results in an overview of main concepts and relations within a domain. We base these steps on what is called open coding (step B), axial coding (step C, D) and selective coding (step D and phase 3). Strauss and Corbin (1990) propose to use open coding to examine and conceptualize field data. During axial coding, researchers reorganize and relate the identified concepts. Finally, during selective coding, researchers select a core concept, refine lower-level concepts and systematically relate and validate conceptual relations. We now describe these steps in greater detail.

Assemble Data

We propose to conduct qualitative data analysis to support knowledge abstraction. To this end, this step involves assembling data that has been collected during phase 1. Qualitative data analysis software such as NVivo (2012) and ATLAS.ti (2010) can support this. The packages enable developers to import and store various types of data in a structured way. In the steps that follow, developers can use this to cross-link various document types, and to subsequently extract concepts, terms, instances and relations.

Explore Main Concepts

To extract the main concepts from the vast amount of assembled data, we propose developers to study the field notes, documents, audio files and pictures that have been integrated in the qualitative data analysis software. While reading passages of text, listening back to fragments of audio and studying pictures; we suggest developers to identify and label the 'pieces of data' that refer to a concept (open coding, Strauss and Corbin 1990). Once each new concept is created, we suggest continuing identifying additional concepts. It is worth noticing that, during this creative step, developers can create as much labels as they can. There is no need to single out potential concepts in this stage. The list can be seen as the first product that conceptualizes the specific domain through the eyes of practitioners.

Organize and Ground All Concepts

By now developers should have a first idea about the main instances, terms and concepts within the domain. To better organize these, we advise to repeat step B once more. This time, however, we suggest using the list of identified concepts as a guide to label data that was not labeled before. While repeating this coding step, developers can also merge, amend and reformulate concepts that were identified earlier.

Visualize Concepts and Relations

In this step, developers establish conceptual and categorical relations (axial coding, Strauss and Corbin 1990) and start refining all relations concepts, terms and instances (selective coding). Qualitative data analysis tools provide features to visually map these relations. With such tools, developers can map out conceptual relations such as: causes, part-of, property-of, is-associated-with and contradicts-with. We suggest visualizing these relations in a tree structure.

Phase 3: Develop and Implement Ontology

The provisional insights from the previous steps allow developers to evaluate the first version of their ontology. In this step we propose developers to convert their concepts and relations in a formal model that represents the ontology in computer interpretable language.

As a next step, we propose to implement the draft formal ontology in an information system. This allows developers to validate the ontology in a real-life setting. In this step, developers introduce their information system to its future application domain. They engage with end-users that use the information system as part of their daily routines. As a first validation step, we propose developers to identify instances of end-users referring to concepts of the draft ontology. Additionally, by observing actual implementation of the information system, developers can identify shortcomings of the information system. These shortcomings feed the identification of additional concepts for the ontology's next version.

All in all, the steps guide developers in gradually creating a valid end-user oriented domain ontology. Although our process model may seem structured in linear phases, we emphasize that several iterations of ethnographic fieldwork, analysis of empirical data and ontology development and implementation are needed to adequately tailor the ontology to end-user needs.

DEVELOPING AN INNER CITY SUBSURFACE UTILITY DOMAIN ONTOLOGY

This section illustrates how we used the fieldwork-based ontology development method to develop an ontology for end-users in the domain of inner city subsurface utility reconstruction projects. We first describe the domain and elaborate on how an ontology could support its construction management information systems. We then describe the conducted steps and conclude by visualizing our informal ontology.

Domain Description

The modeling efforts for this study took place during two case studies of the urban subsurface utility construction project domain. In this domain, a myriad of authorities, service-providers and contractors work simultaneously on interrelated utility construction activities. Separate contractors work on, for example, street interior renewal, reconstruction of sewer-lines, gas and water pipes, electric-grids and fiber-glass cables. To streamline the various construction processes, project managers need to align their construction activities. Information systems can support this challenging coordination task by enhancing managers' understanding of interrelated construction tasks. Additionally, these systems help the managers to create and evaluate alternative construction plans quickly and effectively. Unfortunately, existing information systems lack feasible ontologies that adequately support the end-user viewpoints on inner city utility projects. We therefore conducted two case studies and identified the objects and concepts that professionals use to coordinate utility projects. Based on this, we developed an ontology from scratch.

Process Description

Before we started, we identified our goal and scope: developing an ontology that supports visualizations for the coordination of the inner city subsurface utility construction domain. We limited ourselves to the identification of the main objects that practitioners use to schedule and plan construction projects. We selected two similar multidisciplinary inner city utility projects. Both infrastructure projects took place in a Dutch mid-sized city and comprised reconstruction of a part of the inner city street network. On the first project, one contractor reconstructed an intersection and prepared a connecting main street for future underground construction work. To this end, one main contractor reconstructed street interior and the sewerage-system. Additionally, utility contractors renewed and realigned numerous cables and pipes. Finally, a third contractor drilled sheet piles and removed a large amount of contaminated soil.

The second project involved similar construction work: one contractor was responsible for the reconstruction of the sewerage system and street interior, while another contractor reconstructed freshwater tubes, electricity and fiber glass cables, gas pipes and telecommunication lines. Also in this project, existing subsurface infrastructure had to be replaced to prepare future underground construction work. Finally, a contractor of an adjacent hospital project was involved to align construction works on shared public space.

To kick-off the ontology development, we introduced our research to professionals working on the two case projects. Both the project clients and contractors allowed us to study their projects and invited us for their multidisciplinary coordination meetings. In the first project, we mainly used ethnographic research to explore main concepts from the domain. To this end we attended bi-weekly meetings that took place prior to the actual construction work. We also conducted observations during weekly construction site meetings and additionally observed ongoing construction activities. We tracked our observations as we made field notes. Shortly after each observation event, we processed these notes in a more detailed file. We further audiotaped fourteen meetings, took about three-hundred pictures and collected a few designs and meeting minutes.

During the second project, we collected data similarly; we attended multi stakeholder planning meetings and site visits. Additionally, we developed a draft of our ontology and implemented it in a 4D-CAD information system. The system was implemented in a series of real-life multi-stakeholder coordination meetings. We observed the system implementation to validate our ontological concepts for its end-user context. We summarized characteristics of our fieldwork in Table 7-2 and provide detailed descriptions of our steps below.

Table 7-2. Project characteristics and fieldwork description

	Project 1	*Project 2*
Construction work	Replacing sewerage, water and gas pipes, electricity and data cables, and reconstructing intersection to prepare deeper underground construction work	
Stakeholders	Municipality, service providers, civil works contractor, utility contractor, sheet pile contractor, various subcontractors	Municipality, service providers, civil works contractor, utility contractor, hospital contractor, various subcontractors
Ethnographic steps	• Observed 3 coordination meetings • Observed 4 coordination meetings	• Observed 4 coordination meetings • Conducted 6 work planner interviews • Observed 2 onsite meetings
Action research steps		• Implemented ontology in 4D-CAD model • Evaluated model during 3 additional coordination meetings • Evaluated model during 2 work planner meetings
Audiotaped meetings	14	16
Pictures taken	297	50
Designs and minutes collected	3	8

The vast amount of qualitative data that we collected on the first project allowed us to make a first conceptualization of coordination objects in the utility construction project domain. To do this, we conducted open and axial coding (described in phase two and three of our method). For open coding, we first imported our field notes in the qualitative data coding software ATLAS.ti (2010).

Subsequently, we used this software to mark excerpts from our notes and label each with a string of descriptive text. The following citations show examples of our quoted data (translated from Dutch):

'we have a district station right over there [. . . and . . .] plan to move it over there in the future [. . .]But can't we do this during our current project?' [utility client]

'perhaps we need to replace the gas-case from here to there' [utility client]

'we start moving the gas-case at the beginning of next week' [site manager]

'what about the public lightning? [I mean, do we also install] power-supply cables for public lightning? [site manager] Yes, this will also be incorporated part of your work' [public client]

'the public lightning will also be moved' [public client]

'will we remove the phone booth? I'm not sure, it is not part of the official assignment' [utility client]

'we also have the safety barrier over here. I know, that can also be removed' [contractor]

'do also we need to remove the road signs?' [contractor]

'the container is also located over here, so we have to move it' [public client and contractor]

To abstract these quotes, we grouped them in higher level codes such as *gas district station, lightning cable, lightning column, phone booth, safety barrier, road marking and subsurface waste container*. Subsequently, we used these codes to create a tree-structure that related our codes more explicit. The resulting coarse overview of important concepts, terms and instances helped us to create a first understanding of the utility construction project domain. As these steps increased our domain knowledge and semantic understanding, we could clean the coarse ontology during an axial coding round. In this step, we labeled and coded all data again to provide more detail and find concept relations. For example, our open coding round we derived the concept *gas pipe*. In the successive axial coding round, we identified related lower level concepts and attributes such as *gas pressure, gas pipe material, welding points and pipe protective material*. The resulting list of concepts culminated in our draft ontology.

In a subsequent step, we used the concepts from our draft ontology to guide our ethnographic observations on the second project. Additionally, we conducted action research steps by developing a 4D construction process visualization. We created a project-specific 4D-model incorporating concepts and objects of our ontology. Subsequently, we demonstrated the model during work planner and multi-stakeholder coordination meetings. During these sessions, we validated our draft ontology by observing whether our concepts from the 4D tool were used in discussions. In this coding round, we confirmed earlier identified end-user

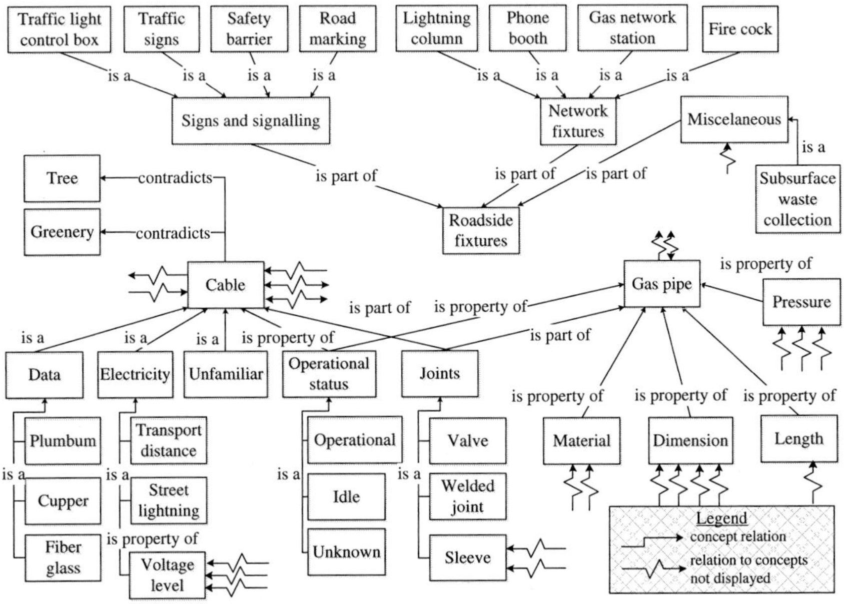

Figure 7-2. Selection of the informal ontology we developed using our fieldwork-based method in two cases

concepts such as *trees, fences* and *welding end-openings*. The observations further provided the additional concept *subsoil*. After two rounds of coding on data from project 1, we conducted selective coding and organized and visualized our draft ontology. Abstractions from project 1 allowed us to further improve our conceptual domain understanding. We used the insights in project 2 to validate and refine concepts. During the 4D implementation, for example, we identified the additional object *foundation block*.

The demonstrated development process continues with various additional iterations between data collection and coding. In essence, we could iterate back and forth continually to refine our domain ontology. As this section is merely illustrative for our method, we decided not to elaborate on these repetitive steps. The final version of our informal ontology comprised 7 high-level concepts, 91 lower-level concepts and 108 concept relations. We depict a selection of this informal draft ontology in Figure 7-2.

DISCUSSION

This study focused on end-user involvement in ontology development methodologies. We analyzed popular methodologies from ontology development and engineering literature and proposed complementary steps for exploration of end-user view on domains. Our fieldwork-based method brings developers closer to the context of the end-user through techniques such as observations, ethnographic interviews and ontology implementation. This inductive conceptualization process allows developers to ground their findings in empirical data. In our field study, for example, the collection of pictures, meeting minutes, audio tapes, ethnographic interviews and observations of meetings allowed us to ground our concepts in perspectives of end-user in the utility domain. The additional action research we conducted by implementing a 4D-CAD information system further allowed us to validate and further increase user-fitness of our utility domain ontology.

We feel that the presented method can be applied for ontology development in two distinct ways: (1) as in our case study, practitioners could use our method to develop ad-hoc ontologies from scratch, using a bottom up approach. Alternatively, (2) developers can use our approach to involve end-users to further validate and complement ontologies that already exist. In this way, our method helps developers to add ad-hoc, or move out redundant, concepts which contextualize ontologies to domain specificities. In our study, for example, we could have used existing object models such as IFC and CityGML as point of departure for our domain ontology. One drawback of this latter approach is, however, that developers need to switch continually between top-down and bottom-up conceptualizing as they try to fit ad-hoc concepts in the existing ontology. It can, for example, be difficult to merge ad-hoc concepts with existing top-down-developed concepts as they seem similar, while they actually imply different meanings. Hence, a sound balance needs to be found between top-down and bottom-up approaches to adequately account for user needs of the ontology. Since we could not reveal much about the boundaries of these two distinct approaches, future research is needed to clarify these boundary issues.

One other limitation of this study is that it cannot yet provide conclusive findings with respect to generalizability of identified ad-hoc concepts. This is problematic especially for ontologies that are developed from scratch. Future research should hence focus on developing criteria and steps for distinguishing ad-hoc concepts from the ones that have a high validity for other projects. Such research additionally positions our research better within conceptual modeling debates on reliability of using observational data, reasoning, capturing details, unique views and model generalizability.

Finally, our method assumes frequent stakeholder interaction during concept exploration, validation and evaluation phase. Although this clearly provides benefits with regards to user-fitness and feasibility, we know little about related development costs and resources. Future research should hence compare costs of our approach against traditional post-development methods for user-involvement.

CONCLUSION

This study argues that, to be successfully implemented, information systems need ontologies that integrate various end-user viewpoints. Ontologies that fail to do this are of little relevance for practitioners and hamper the adoption of an information system. This chapter shows that various existing ontology development methodologies contain little description of how they aim to address this issue. Current methods for data acquisition, conceptualization and validation suggest, for example, end-user based scope definitions, brainstorm sessions and post development interviews with domain experts. Unfortunately, such steps are often not spelled out formally and limitedly succeed in capturing or verbalizing end-user knowledge. This chapter therefore extends current top-down ontology development approaches and introduces ethnographic-action research. We developed a formal method supporting the creation of ontologies that fit the end-user viewpoints and demonstrated the approach for the domain ontology of the inner city utility construction projects.

Using our method—on its own or in addition to existing methodologies—will allow developers to enhance the user-fitness and representativeness of their domain ontologies. This subsequently customizes information systems more towards end-user needs. In future, this is likely to contribute to more effective information system implementation and adoption processes.

ACKNOWLEDGEMENTS

We greatly acknowledge Stichting Pioneering for subsidizing this research. We are also thankful to Gemeente Enschede and its project stakeholders for granting us full-access to research their two projects.

References

Agar, M. H. (1996). *The professional stranger: An informal introduction to ethnography*, Academic Press, San Diego.

ATLAS.ti. Version 6.1. [computer software] (2010). Berlin, Scientific Software Development GmbH.

Baskerville, R. L. (1999). "Investigating information systems with action research." *Communications of the AIS*, 2(19), 2–31.

Björk, B. C. (1992). "A unified approach for modelling construction information." *Building and Environment*, 27(2), 173–194.

Chandrasekaran, B., Josephson, J. R., et al. (1999). "What are ontologies, and why do we need them?" *Intelligent Systems and Their Applications, IEEE*, 14(1), 20–26.

El-Diraby, T. (2012). "Epistemology of Construction Informatics." *Journal of Construction Engineering and Management*, 138, 53–65.

Fernández-López, M. (1999). "Overview of methodologies for building ontologies." in *Proceedings of the IJCAI-99 workshop on Ontologies and Problem-Solving Methods (KRR5)*, Stockholm, Sweden.

Gómez-Pérez, A., Fernández, M., and de Vincente, A. J. (1996). "*Towards a method to conceptualize domain ontologies*."

Gómez-Pérez, A., Fernández-López, M., and Corcho, O. (2004). *Ontological Engineering: With Examples From the Areas of Knowledge Management, E-commerce and the Semantic Web*. Springer, London.

Gruninger, M. and Fox, M. S. (1995). "Methodology for the Design and Evaluation of Ontologies."

Guarino, N., Oberle, D., and Staab, S. (2009). "What is an Ontology?" in *Handbook on Ontologies*, eds Staab, S. and Studer, R., Springer-Verlag, Berlin Heidelberg, 1–17.

Hartmann, T., Fischer, M., and Haymaker, J. (2009). "Implementing information systems with project teams using ethnographic-action research." *Advanced Engineering Informatics*, 23(1), 57–67.

Hsieh, S. H., Lin, H. T., Chi, N. W., et al. (2011). "Enabling the development of base domain ontology through extraction of knowledge from engineering domain handbooks." *Advanced Engineering Informatics*, 25(2), 288–296.

Jorgensen, D. L. (1989). *Participant observation: A methodology for human studies*, Sage Publications Inc. Thousand Oaks.

Katranuschkov, P., Gehre, A., and Scherer, R. J. (2003). "An ontology framework to access IFC model data." *ITcon* Vol. 8, 413–437, retrieved from http://www.itcon.org/2003/29 on Aug. 26, 2013.

Kishore, R., Sharman, R., and Ramesh, R. (2004). "Computational ontologies and information systems: I. foundations." *Communications of the Association for Information Systems*, 14, 158–183.

López, M. F., A., Gómez-Pérez, and Sierra, J. P., et al. (1999). "Building a chemical ontology using methontology and the ontology design environment." *Intelligent Systems and Their Applications, IEEE*, 14(1), 37–46.

NVivo qualitative data analysis software [computer software]; QSR International Pty Ltd. Version 10, 2012.

Phelps, A. F. and Horman, M. J. (2010). "Ethnographic theory-building research in construction." *Journal of Construction Engineering and Management* 136, 58–65.

Pinto, H. S., Tempich, C., and Staab, S. (2009). "Ontology engineering and evolution in a distributed world using DILIGENT." in *Handbook on Ontologies*. Eds Staab, S. and Studer, R. Springer-Verlag, Berlin Heidelberg, 153–176.

Russell, S. J., Norvig, P., Canny, J. F., et al. (1995). *Artificial intelligence: a modern approach*, Prentice Hall Englewood Cliffs, NJ.

Simon, H. A. (1976). *Administrative behavior—a study of decision making processes in administrative organisations*, Cambridge University Press.

Spradley, J. (1979). *The ethnographic interview*, Holt, Rinehart and Winston, NY.

Strauss, A. L. and Corbin, J. M. (1990). *Basics of qualitative research: Grounded Theory Procedures and Techniques*, Sage Publications, Inc. Newbury Park.

Studer, R., Benjamins, V. R., and Fensel, F. (1998). "Knowledge engineering: principles and methods." *Data & knowledge engineering*, 25(1), 161–197.

Sure, Y., Staab, S., and Studer, R. (2009). "Ontology Engineering Methodology." in *Handbook on Ontologies*. eds. Staab, S. and Studer, R., Springer-Verlag, Berlin Heidelberg, 135–152.

Turk, Z. (2001). "Phenomenologial foundations of conceptual product modelling in architecture, engineering and construction." *Artificial intelligence in engineering*, 15(2), 83–92.

Uschold, M. (1996). "Building ontologies: Towards a unified methodology." in *Proceedings of Expert Systems '96, the 16th Annual Conference of the British Computer Society Specialist Group*, University of Edinburgh, Cambridge.

Uschold, M. and M., Gruninger (1996). "Ontologies: Principles, methods and applications." *Knowledge engineering review*, 11(2), 93–136.

Yin, R. K. (2003). *Case study research: design and methods*. Sage Publications, Thousand Oaks, CA.

CHAPTER 8

Transaction Formalization in Infrastructure Management Using an Ontological Approach

J. Zeb*
T. Froese†

Abstract: *Municipal organizations use a range of information systems to manage their infrastructure systems. Issues associated with these information systems include the fact that the underlying data formats' heterogeneity cause interoperability problems; and data are commonly exchanged between various agencies in a manual and ad hoc way. The growing trend is to transform the current practice of manual data exchange to a more formalized computer-to-computer-based information exchange. For computers to talk to each, the data exchanges need to be defined in a computer interpretable format—the ontology.*

This chapter presents the detailed development and evaluation of two ontologies and briefly introduce their application. The Transaction Domain Ontology was developed to represent knowledge about information transactions to support the design, management and implementation of transactions in the area of infrastructure management. The Tangible Capital Asset Ontology was created to represent the physical components that make up infrastructure systems to support the design of messages used to report the inventory and condition of infrastructure assets. A ten-step approach was devised to develop both the ontologies following a layered architecture. As part of the evaluation, both the ontologies were verified and validated.

---

*Department of Civil Engineering, The University of British Columbia, Civil and Mechanical Engineering, 2002-6250 Applied Science Lane, Vancouver, British Columbia, Canada, V6T 1Z4; PH (604) 822-2739; FAX (604) 822-6901; email: jzmir1@interchange.ubc.ca

†Department of Civil Engineering, The University of British Columbia, Civil and Mechanical Engineering, 2002-6250 Applied Science Lane, Vancouver, British Columbia, Canada, V6T 1Z4; PH (604) 822-2739; FAX (604) 822-6901; email: tfroese@civil.ubc.ca

INTRODUCTION

Municipal infrastructure organizations own, operate, and manage infrastructure systems to provide un-interrupted service to various communities. Infrastructure organizations use diversified information systems to manage infrastructure systems. Problems pertaining to these information systems include; heterogeneity of the underlying data format and the manual (human-to-human) and ad hoc data exchange between infrastructure organizations. These problems hamper interoperability between information systems of infrastructure organizations.

Some examples of information exchange between infrastructure organizations include: communications during disaster response (e.g. Is power available in this area? Who is responsible for this section of roadway? When will water be restored to this area?); coordination between buried utility agencies to provide a 'call-before-you-dig' call center for excavations; or aggregating data from multiple infrastructure management software for the purpose of generating comprehensive municipal infrastructure condition reports to meet public sector accounting requirements. An element of the infrastructure reporting from this last example that has been identified is an Asset Inventory and Condition Assessment Reporting or Tangible Capital Asset Reporting (AI&CAR/TCA Reporting) (Zeb et al. 2012). In this transaction, different municipalities report their tangible capital asset information to the provincial government for financial planning in order to fulfill the Public Sector Accounting Board reporting requirements, (PSAB, 2009). According to Felio (2012), infrastructure management organizations find it difficult to exchange the tangible capital asset information due to heterogeneity of data; lack of consistency in class description; and lack of aggregation of the data. The growing trend is to formalize these transactions and transform the current practice of manual data exchange to a more formalized computer-to-computer data exchange. The issue is how to formalize transactions for computer-based data exchange.

An ontology-supported transaction formalism protocol was developed to address this issue. The protocol is an eight–step procedure developed to formalize transactions in the domain of infrastructure management. These steps are: assess need, define an as-is transaction map, develop a to-be transaction map, collect information, design the message template, review transaction map and message template, adopt and implement a standard transaction agreement/transaction specification (the combination of the transaction map and message templates), and monitor the standard transaction agreement/transaction specification. The overall approach used to develop and apply the proposed protocol at three levels of abstraction is presented in Figure 8-1, which shows the actors involved, process/tool used, and output obtained. The three levels of abstraction are: (i) developing the protocol, (ii) designing the transaction, and (iii) performing the transaction.

i. Develop the protocol: the researcher designed a conceptual model of the proposed protocol that is referred to as a transaction formalism protocol specification. The proposed protocol allows a transaction designer to formalize a transaction in terms of transaction maps (sets of sequenced atomic

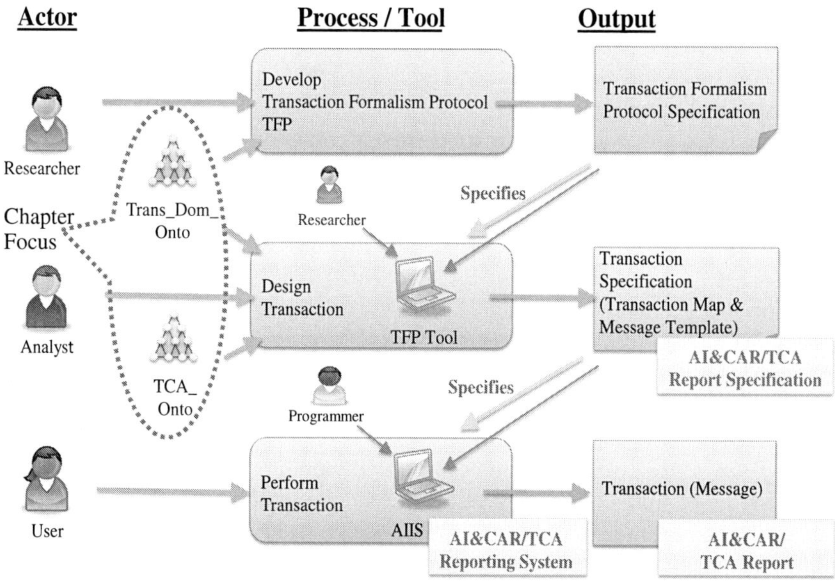

Figure 8-1. Research abstraction layers

transactions) and message templates (representing header and payload information in a structured format). The protocol draws upon the Transaction Domain Ontology (Trans_Dom_Onto) for elements that make up the design, implementation, and management of the transaction map and message template header information (meta information about a message template).

ii. Design the transaction: the transaction formalism protocol tool implements the protocol specification in the form of a software application. The protocol tool comprises of a set of forms developed to create standard transaction agreements/transaction specifications. The researcher developed the transaction formalism protocol tool, which would then be used by an "analyst" to define transactions that end users will use. In this research work, the tool was applied to formalize and create the AI&CAR/TCA Reporting specification. For the message template payload information (the body of the information that is exchanged in a transaction); the protocol uses the terms represented in different ontologies and data models depending upon the area of application. For the AI&CAR/TCA Reporting transaction, the payload information was captured from the Tangible Capital Asset Ontology (TCA_Onto) developed as part of this research. The message header information (meta information about the message and transaction) was captured from the Trans_Dom_Onto.

iii. Perform the transaction: at this level, an end user will carry out individual transactions in accordance with the transaction specification. Often, the transaction specification will be implemented into a piece of software by the software developers (programmers), and the end user need have no knowledge of the transaction specification. As part of this research, the AI&CAR/TCA Reporting specification was developed that was then implemented in a prototype Asset Information Integrator System. This is a web-based prototype reporting system that is to be used by different municipalities to report their tangible capital asset information to the provincial government for financial planning and budget allocation.

The advantages of the proposed solution are to enable the transaction development personnel (transaction analysts, transaction designers, software developers, process modellers, and industry experts) to develop standard transaction agreements/transaction specifications effectively and efficiently; and define the transaction specifications consistently to accomplish message-based interoperability between information systems of the infrastructure organizations.

The core focus of this chapter is on the development and evaluation of the Trans_Dom_Onto and TCA_Onto in the domain of infrastructure management. This chapter comprises of nine sections. The *first section* identifies the problem, background information, and the proposed solution. The *second section* describes related research work in the area of ontology development. The *third section* explains the methodology used to develop an ontology while the *fourth section* discusses the ontology architecture. The *fifth and sixth section* represents the development of the Trans_Dom_Onto and the TCA_Onto respectively. The *seventh section* introduces the the ontology application and *eighth section* discusses the evaluation. Finally, the *ninth section* describes the research conclusions.

RELATED WORK IN ONTOLOGY DEVELOPMENT

Three ontologies that are of particular importance for this research have already been developed in the domain of infrastructure management. These ontologies are the Infrastructure Product Ontology (Osman, 2007) representing infrastructure products; the Infrastructure and Construction Process Ontology (El-Gohary, 2008) representing various processes over the life cycle of projects; and the Actor Ontology (Zhang and El-Diraby, 2009) representing the actors playing diversified roles within the construction industry. These ontologies represent a range of infrastructure domain knowledge; however, lack the knowledge related to transaction design, management and implementation in the domain infrastructure management, e.g. transactions, actor roles, messages, communication channels, etc.

The Open-electronic Data Interchange Transaction Ontology, (ISO, 2006) is another related ontology that focuses on the design and management of commercial transactions (buy/sell transactions). This ontology evolved from the resource-event-agent ontology (Allen and March, 2006) that was based on an accounting model resource-event-actor (McCarthy, 1982). The knowledge in the Open-electronic Data Interchange Transaction Ontology is modelled from three perspectives: financial (exchange something of value), commercial (business markets) and industrial (business location) with specific focus commercial transactions, whereas the emphasis of this research work is on information transactions in the domain of infrastructure management. This requires the development of a Trans_Dom_Onto to represent knowledge related to design, management and implementation of transactions and message template header information.

The knowledge relating to transaction design and transaction headers is fairly standard across a range of all formal transactions for any domain, but there are elements that are tailored to the domain of infrastructure management (e.g. the specific types of actor roles that would be found within the infrastructure industry). In contrast, the knowledge relating to the information content (payload) of transactions is very dependent on the domain and the specific context. For the design of transaction payload information, a number of different ontologies and data models can be used depending on a specific area of application. For the design of message templates for the AI&CAR/TCA Reporting transaction, the previously mentioned Infrastructure Product Ontology can be used; but some extensions are required. To fill this gap, the TCA_Onto was developed as part of this research work as an extension to the Infrastructure Product Ontology. In the TCA_Onto, infrastructure products are represented in four categories: transportation, water, wastewater, and solid waste management. According to PSAB (2009), tangible capital assets are "non-financial assets having physical substance that are acquired, constructed or developed and: are held for use in the production or supply of goods and services; have useful lives extending beyond an accounting period; are intended to be used on a continuing basis, and are not intended for sale in the ordinary course of operations."

ONTOLOGY DEVELOPMENT METHODOLOGY

To develop both the Trans_Dom_Onto and TCA_Onto, a ten-step methodology was devised as shown in Figure 8-2, which is a hybrid version of the methodologies developed by: (i) Gruninger and Fox (1995); (ii) Uschold and Gruininger (1996); (iii) Fernandez-Lopez et al. (1997); and (iv) Noy and McGuinness (2001). A brief description of each step is listed below.

i. Define ontology coverage—the purpose, usability, and scope of the ontology was defined.
ii. Capture competency questions—a set of competency questions was defined based on the requirement analysis. According to Gruninger and Fox (1995), competency questions represent a set of requirements that the ontology should be able to answer.
iii. Create taxonomy—a 4C approach was used to develop taxonomies of concepts. These steps are: capture, compare, categorize, and create taxonomies.
iv. Reuse and merge existing ontologies–where possible, relevant existing ontologies were used.

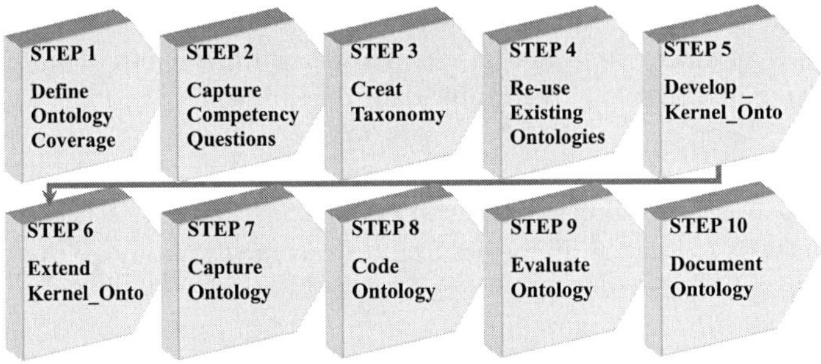

Figure 8-2. Ten-step methodology to develop ontologies

v. Develop kernel ontology—a kernel ontology was first developed that represents transaction domain knowledge and tangible capital asset knowledge (concepts) at an abstract level.

vi. Extend kernel ontology—each concept was extended to create detailed taxonomies.

vii. Capture ontology—involves the development of axioms. According to Gruninger and Fox (1995), axioms describe a concept unambiguously and constrains its' interpretation.

viii. Code ontology—the knowledge was formally coded using the Web Ontology Language in an open-source Protégé ontology editor (Protégé, 2014).

ix. Evaluate ontology—both the ontologies were evaluated as explained in the evaluation section.

x. Document ontology—finally, the knowledge representation was documented for future use.

A set of the following constraints and difficulties are associated with the application of the research methodology.

i. It is difficult to formulate a set of meaningful competency questions that the ontologies is to answer.

ii. Identification of the relevant concepts and its' categorization into modality-based taxonomies is an iterative, cumbersome and time-consuming process.

iii. In case an existing relevant ontology is to use, its' identification and evaluation is difficult as it requires specific expertise in ontology evaluation and ontology merging.

iv. The development of the hard axioms (concept declaration using formal language) in the ontology requires skills in applying first order logic.

v. The add-in visualization applications in the Protégé ontology editor limits visualization of the knowledge representations to "is-a" relationships only; therefore, to show other than is-as relationships (i.e. directed association relationships), a graphical representation tool, e.g. Unified Modeling Language or other similar tool is required.

vi. For ontology evaluation, identification and selection of industry experts is a real challenge.

ONTOLOGY ARCHITECTURE

An ontology is "an explicit formal specification of the terms in the domain and the relations among them" (Gruber, 1995). According to Gomez-Perez et al. (2005), ontologies are built in a layered architecture as shown in Figure 8-3 where each layer represents a specific level of abstraction of a conceptualization. Based on different levels of abstraction, ontologies are of four types as described below. The boxes with a gray background in Figure 8-3 shows original work developed as part of this research.

- Upper ontology: represents the knowledge in a domain of interest at a very high level of abstraction where concepts are the most generic and common across different industries. The Transaction Upper Ontology (level-1) represents the transaction knowledge at the most abstract level that is organized according to core and support concepts. The *core concepts* are action (i.e. process and event), product, project, resources (human, physical, financial, and information), whereas the *support concepts* include modeling concepts (modality and attribute), mechanism, constraint, and relationship as defined in Zeb and Froese 2011.
- Domain ontology: represents the knowledge related to a specific domain of interest that can be used to create different application-level ontologies and may be shared by different applications. The Trans_Dom_Onto (level-2) was created at two levels of abstraction. The *Transaction Domain Kernel Ontology* (level 2-1) represents the knowledge at the abstract level with fewer concepts organized according to core concepts (e.g. transaction, message, message instance, actor/actor role, and information) and support concepts (e.g. transaction communication channel, transaction modality, transaction constraint, and relationship as defined in Zeb and Froese (2012). The *Transaction Domain Extended Ontology* (level 2-2), represents detailed taxonomies of the core and supports concepts. To integrate the infrastructure domain knowledge, a link was established between the Trans_Dom_Onto and other existing infrastructure domain ontologies: Infrastructure and Construction Process Ontology, Actor Ontology, and Infrastructure Product Ontology.
- Application ontology: represents the knowledge required to develop a specific application. The transaction application ontology—the TCA_Onto (level-3) was developed as an extension to the Infrastructure Product Ontology to support the design of message templates for the AI&CAR/TCA Reporting. The TCA_Onto was created at two levels of abstraction. The *Tangible Capital*

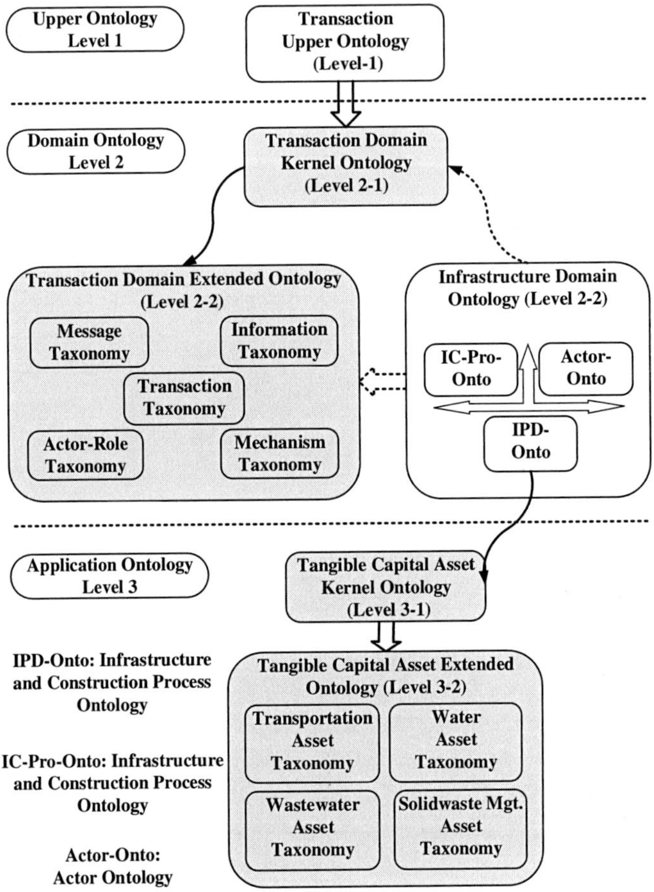

Figure 8-3. Ontology architecture

Asset Kernel Ontology (level 3-1) represents four main modalities of the tangible capital assets at the abstract level. The *Tangible Capital Asset Extended Ontology* (level 3-2) represents sector-based detailed taxonomies of the tangible capital assets.

- User ontology: represents knowledge captured by a specific user of an application. The user ontology was not developed in this research work.

TRANSACTION DOMAIN ONTOLOGY

This section describes taxonomies of the concepts represented in the Transaction Domain Kernel Ontology. These concepts were organized according to core concepts (transaction, message, actor/actor role and information) and support concepts (mechanism, modality, attribute, axiom, constraint and relationship).

Taxonomies of core and support concepts were developed according to the concept of modality. According to El-Gohary (2008), modality is a "characteristic that describes a thing and denotes it's belonging to a particular group or category." These modalities are created based on the review of the related ontologies, data models and discussion with the domain experts. This section briefly describes the modality-based taxonomy development of the core and support concepts at the very abstract level.

Transaction Taxonomy

A transaction is defined as any communication between the sender and receiver roles that results in the flow or exchange of information through a single or a sequenced set of messages. Transactions are classified based upon the concept of modality. The two main transaction modalities are: communication transaction-modality and domain transaction-modality (as illustrated in Figures 4 and 5, see Zeb and Froese (2012), for a more complete explanation of these modalities).

Communication Transaction-Modality

The communication transaction-modality classifies transactions based on the way transactions are communicated between the sender and receiver role. It has the following seven sub-classes as shown in Figure 8-4:

- Pattern transaction-modality: classifies transactions based on the interaction patterns between the actor roles. For example: (i) one-action with and without acknowledgement; (ii) two-action with and without acknowledgement; and (iii) information without acknowledgement. The transaction map for the AI&CAR/TCA Reporting was designed based on one-action and two-action design patterns between different actor role.
- Business transaction-modality: categorizes transactions based upon the type of resource that flows between the sender and receiver role as a result of any communication or interaction between them. These transactions are: commercial transaction (money for goods), financial transaction (money for money) and information transaction (information for information).

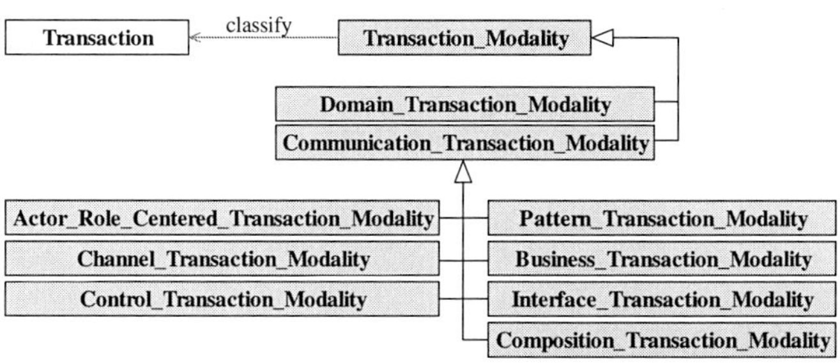

Figure 8-4. Communication transaction-modality, (transaction taxonomy)

- Interface transaction-modality: classifies transactions based on the organizational boundaries through which the sender and receiver roles exchange information. A transaction is said to be internal if the information is exchanged within a single organization and is external if the information is exchanged between two or more organizations.
- Composition transaction-modality: classifies transactions based on its' composition. A transaction is said to be atomic (if a conversation between roles completes in a single communication), compound (if a conversation between roles completes in more than a single communication using any design pattern), or composite (if a communication between roles completes in more than a single communication using a set of compound transactions).
- Actor role centered transaction-modality: categorizes transactions based on the interaction, location, and response timings of the sender and receiver roles in a given transaction.
- Channel transaction-modality: classifies transactions based upon the mode or media through which information is exchanged between the parties in a given transaction. For example, e-mail, fax, postal, etc.
- Control transaction-modality: views a transaction based upon the control on transaction transmission and transaction access.

Domain Transaction-Modality

The domain transaction-modality categorizes transactions based on a specific civil engineering field to which a it belongs. It has four sub-classes as shown in Figure 8-5:

- Sector transaction-modality: classifies transactions based on the engineering sector to which it belongs. For instance, transportatation, water, wastewater, solid waste management, gas, telecom, electricity, building/facility sector transactions.
- Project service delivery-modality: categorizes transactions based upon the mode of project service delivery. For example, design-bid-build, design-build, design-build-operate, design- build-operate- finance, and construction management transactions. This categorization is important because for the same request for information transaction, actor roles could require different information in different project service delivery modes.

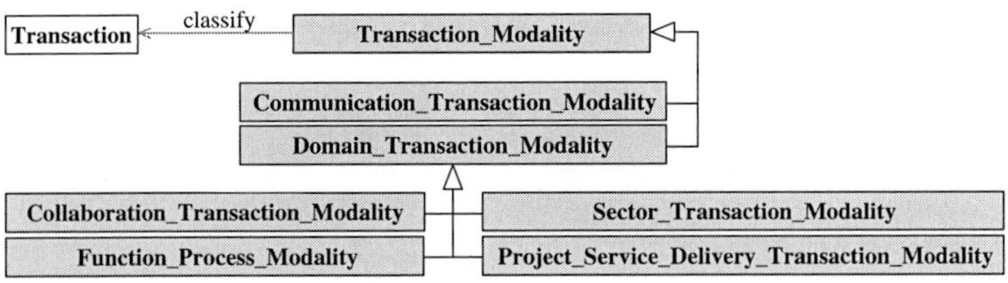

Figure 8-5. Domain transaction-modality, (transaction taxonomy)

- Function process-modality: classifies processes based on the function it performs during the life cycle of the infrastructure project, (El-Gohary, 2008). The function based processes are: core, management, knowledge integration, and support processes. The support processes can further be classified as information management, administrative, and communication management processes. Communication management processes are of particular interest to this research work, which focuses on the exchange of information between collaboration partners. For the design and management of transactions, function processes (specifically the communication management processes) will be captured from the Infrastructure and Construction Process Ontology; therefore, a link of the Trans_Dom_Onto was established with it.
- Collaboration transaction-modality: categorizes transactions based upon the number of collaborating parties involved to complete a transaction.

Message Taxonomy

A transaction consists of three elements: transaction map, actor role and message/message template. Therefore, a transaction message is part of a transaction. According to Zeb and Froese (2012), a transaction message refers to the "information in tangible (written) and intangible (verbal) forms that is exchanged between the parties in a given transaction." The message modality classifies different types of messages in the domain of infrastructure management. It has four sub-classes as shown in Figure 8-6:

- Function message-modality: classifies messages based upon the type of function it performs in a given transaction.
- Formulation message-modality: categorizes messages based the way a message is created or formulated, i.e. a verbal (representing intangible information) or written (representing tangible information) message.
- Representation message-modality: classifies messages based upon the degree to which information is structured in a message.
- Intelligent message-modality: classifies transactions based on the level to which a message is computer interpretable.

Actor and Actor role Taxonomy

According to ISO (2006), an actor is either an individual or an organization. An *individual* is a human-being and is indivisible, whereas an *organization* is a

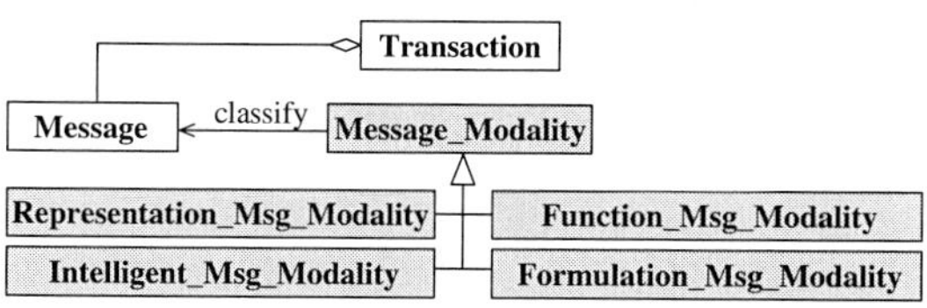

Figure 8-6. Modality-based message taxonomy

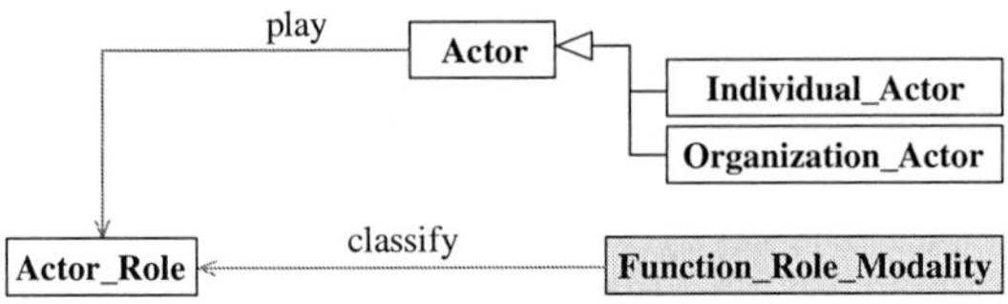

Figure 8-7. Modality-based actor and actor role taxonomy

framework of authority where actors play certain roles to achieve a common goal. An actor plays a variety of roles that is classified according to the function role modality as shown in Figure 8-7:

- Function role-modality: classifies actor roles based on the function that the actors play in the domain of infrastructure management. According to Zhang and El-Diraby (2009), *actor roles* are "a set of connected behaviors and attributes as conceptualized by actors in a given social position."

Information Taxonomy

A transaction is successfully accomplished once information is exchanged or transferred between the parties involved in the communication. Information is an important element of a transaction. According to OCCS (2006), information is defined as "data referenced and utilized during the process of creating and sustaining the built environment." An information modality classifies the information represented in a message template. It has two sub-classes as shown in Figure 8-8.

- Header information-modality: classifies message header information (i.e. a meta information about a transaction or message) that is represented in a message template. According to RosettaNet (2002), header information are of three types: preamble, delivery and service header information.
- Payload information-modality: classifies message payload information (i.e. actual information content that collaboration parties require to exchange in a given transaction) based on the way that: (i) information is placed in a message; (ii) information is created or formulated; and (iii) information is delivered to other parties.

Transaction Attribute Taxonomy

According to Osman (2007), an attribute is a characteristic, feature, or property that describes a thing, entity, or concept. A transaction has the following set of transaction attributes as shown in Figure 8-9:

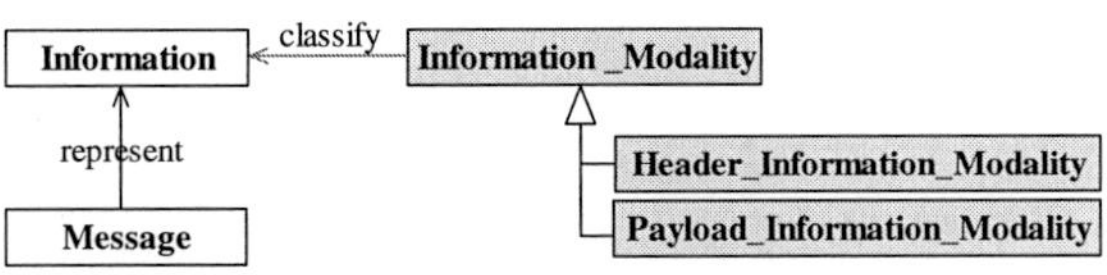

Figure 8-8. Modality-based information taxonomy

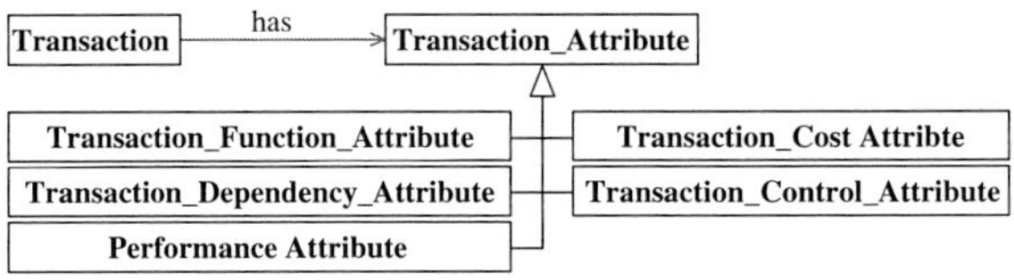

Figure 8-9. Transaction attribute taxonomy

- Transaction function attribute: is a characteristic that describes a transaction based upon the function it performs in a given communication. A function can be to disseminate information or request the receiver to perform an action.
- Transaction dependency attribute: is a characteristic that describes a transaction based on the logical, geographic, and cyber dependency.
- Transaction performance attribute: describe the performance of a transaction in terms of its transaction efficiency. A transaction is said to be efficient if it is effective in terms of time, costand quality.
- Transaction cost attribute: describes the transaction in terms of design, implementation and operational cost.
- Transaction control attribute: describes the transaction security in terms of transaction authorization and authentication.

TANGIBLE CAPITAL ASSET KERNEL ONTOLOGY

To support the design of message templates for the AI&CAR/TCA Reporting transaction, the TCA_Onto was developed as part of this research work. The Tangible Capital Asset Kernel Ontology represents the tangible capital assets was developed first, which was further specialized and extended to develop detailed taxonomies of the tangible capital assets in the four infrastructure sectors of transportation, water, wastewater, and solid waste management (making up the Tangible Capital Asset Extended Ontology). This chapter briefly describe the Tangible Capital Asset Kernel Ontology, which represents four modalities of the tangible capital asset knowledge at a very abstract level as shown in Figure 8-10.

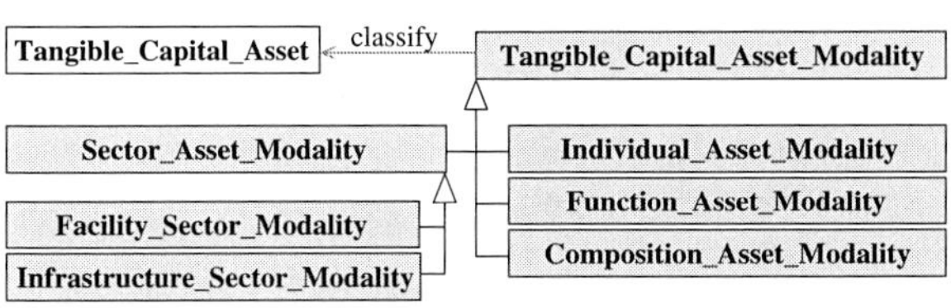

Figure 8-10. Modality-based tangible capital asset kernel ontology

Tangible Capital Asset Taxonomy

The tangible capital asset modality classifies the tangible capital assets based on the following four views:

- Individual asset-modality: classifies the tangible capital assets based on an individual asset type. According to PSAB (2009), there are eight types of the individual assets; land, land improvement, building, infrastructure, machinery and equipment, vehicle, and work in progress.
- Function asset-modality: categorizes the tangible capital assets based upon the function they perform in an infrastructure system. Osman (2007) has identified six types of assets: conveyance, control, access, protection, measuring and storage. Three additional types were identified as part of this research work: filtering, pumping and commuting.
- Composition asset-modality: classifies the tangible capital assets based on their composition in an infrastructure system. Osman (2007) identified and defined three types of function based assets: systems level, sub-system level, and component level.
- Sector asset-modality: classifies the tangible capital assets based on the civil engineering sector to which they belong. This modality has two types. *Facility sector modality*—classifies the tangible capital assets based on the different types of facilities in the construction industry. *Infrastructure sector modality*—classifies the tangible capital assets based on the infrastructure sector to which they belong.

Tangible Capital Asset Attribute Taxonomy

The tangible capital assets have a set of attributes as shown in Figure 8-11. Osman (2007) identified the following attributes for infrastructure products (referred to as tangible capital assets in this research work): cost, dimension, performance, state of operation, dependency, impact, redundancy, spatial, shape and material attributes. Three attributes were added as part of this research work: condition (representing the condition index in terms of very good, good, fair, poor, and very poor for physical and capacity condition); quantitative (quantity and

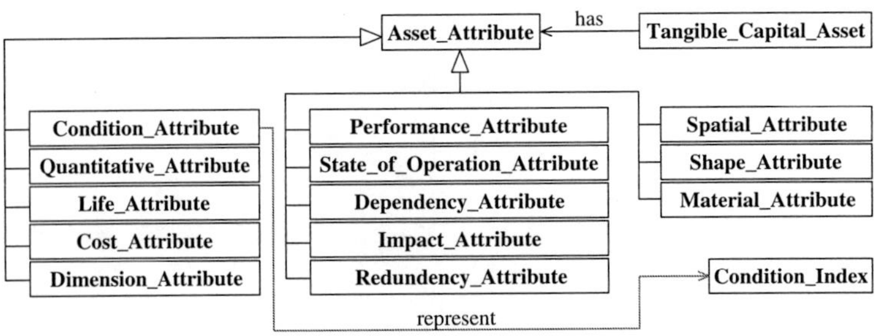

Figure 8-11. Tangible capital asset attribute

disposed quantity); and life (average life, remaining life, and installation year) attributes.

ONTOLOGY APPLICATION

Both ontologies were used simultaneously to define transactions and message templates in the domain of infrastructure management. The AI&CAR/TCA Reporting was identified as one of the transactions that has the greatest potential for information system improvement. In this transaction, municipal governments report information about their tangible capital assets to the provincial government. Currently, the provincial government finds it difficult to compile, compare, and analyze the tangible capital asset information received from several different municipalities due to the heterogeneity of the underlying data format and manual communication through word or PDF reports. These problems are tackled through the development of the ontologies in the domain of infrastructure management to support the design of transaction and message templates that were implemented in a prototype Asset Information Integrator System developed as part of this research. This system collects, compiles, integrates, compares and analyzes the information received from municipalities.

The message template for the AI&CAR/TCA Reporting was defined in eight views. A view is an integral part of a message template that represents the tangible capital asset information related to a specific sector. The first view of the message template represents both the header and payload information while the remaining views show only payload information. Figure 8-12 (a) represents the header information and Figure 8-12 (b) represents the payload information for view 1 of the message template defined for the AI&CAR/TCA Reporting. The header information and payload information fields (related to tangible capital assets in the facility sector) were filled with dummy values.

The following is a description of the potential uses of both the ontologies. The Trans_Dom_Onto was applied from three different perspectives: design, management and implementation.

- Design perspective: views the Trans_Dom_Onto as a dictionary of terms that the transaction development personnel will use to define transactions throughout the eight steps of the proposed transaction formalism protocol tool. The Trans_Dom_Onto explicitly defines the terms so that all parties have a common understanding of these terms and use them consistently. The information represented in the forms developed for each step of the transaction formalism protocol tool was based on the knowledge represented in the Trans_Dom_Onto.
- Management perspective: focuses on how to manage transaction transaction specifications (i.e. standard transaction agreements) over the web. Once transaction specifications are created using the transaction formalism protocol tool, the next step is to archive or store it for future use as part

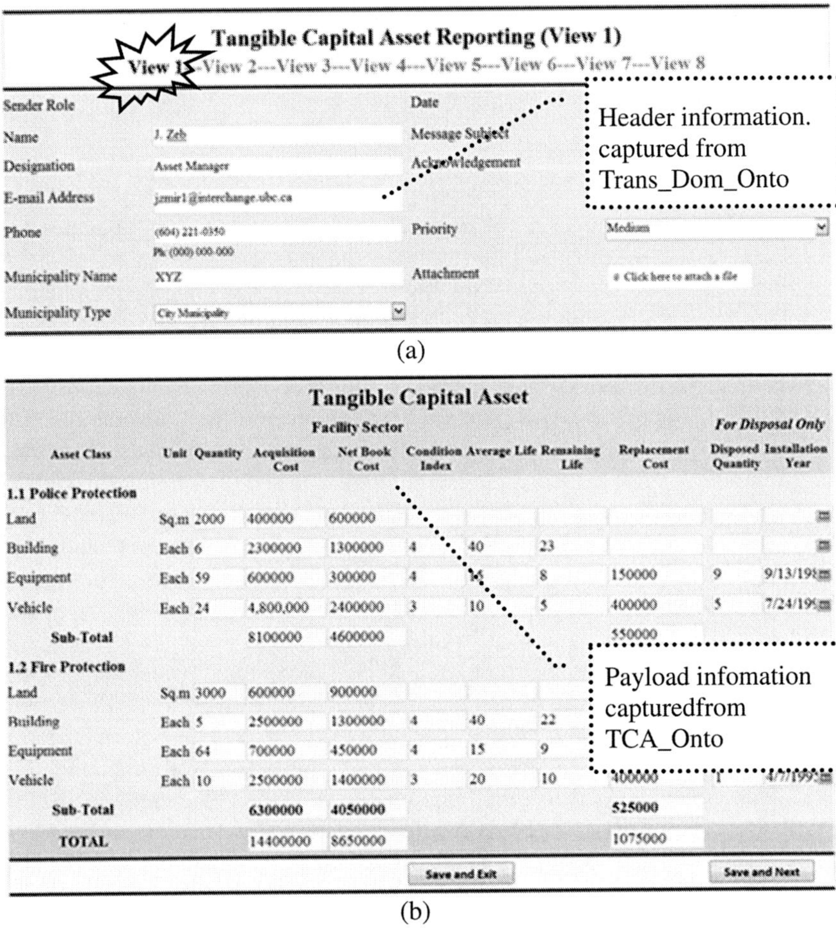

Figure 8-12. Formalized message template–(view 1); (a) Header information, (b) Payload information

of the transaction management. To ease access and retrieval of the transaction specifications, these were stored in a web-based repository—the infrastructure transaction management portal. The portal was constructed based on the classes of transactions defined in the Trans_Dom_Onto. The development of the portal is beyond the scope of this chapter.

- Implementation perspective: the terms represented in the Trans_Dom_Onto can be used to define header part of the message templates. The header information in the message templates developed for the AI&CAR/TCA Reporting was captured from the Trans_Dom_Onto as shown in Figure 8-12(a).

The TCA_Onto, on the other hand, is an application ontology, and the tangible capital asset information represented in the ontology was used to define the payload information part of the message templates developed for the AI&CAR/TCA Reporting as shown in Figure 8-12(b).

ONTOLOGY EVALUATION

According to Gomez-Perez (2001), ontology evaluation is judging the content of the ontology with respect to a frame of reference characterized by a set of requirements, competency questions, and the real-world model of the domain of interest. Ontology verification consists of checking the content of the ontology with respect to a set of modeling requirements and a set of competency questions. On the other hand, the ontology validation involves judging the content with respect to a real-world model through domain experts. A criteria-based approach was used to evaluate both the ontologies following the ontology evaluation framework shown in Table 8-1.

The framework shows the criteria, measures and tools to evaluate both the ontologies. Gomez-Perez (1996) identifies and defines consistency, conciseness, and completeness, Guarino (1998) defines correctness, and Yu et al. (2007) specifies clarity criteria to evaluate ontologies.

- Consistency: refers to the level to which the knolwdge represented in the ontology is consistent, i.e. contradictory conclusions cannot be drawn from the definitions of the concepts represented in the ontology. Three measures were used to check the level to which the criterion was fulfilled. These measures were: circulatory errors (which occur when a class is defined as a sub or super class of itself), partition errors (which occur when disjointedness is omitted), and semantic inconsistency errors (which occur when the class definition is semantically inconsistent within the context of class hierarchy).
- Conciseness: measures the level to which the knowledge represented in the ontology is concise, i.e. no redundant concepts are represented in the ontology. Two measures were used to check conciseness: grammatical redundancy errors (which occur when more than one relationship is defined between the same generalization-specialization classes), and identical formal definition of some class errors (which occurs when two or more classes have the same definition and are represented with different names).
- Completeness: focuses on how complete the knowledge representation is. No measures are developed as yet to measure the completeness of ontology; it is measured in terms of incompleteness in the knowledge representation (Yu et al. 2007).
- Correctness: focuses on how accurately the knowledge representation is modeled from a real- world perspective. Correctness was measured in terms of identity errors (which occur when certain classes in the ontology are not in semantic compliance with the real-world context).

Table 8-1. Ontology evaluation framework

Ontology Evaluation Framework				
		Ontology Evaluation Tools		
		Verification		*Validation*
Criteria	*Measure*	*Automated Reasoner*	*Competency Questions*	*Expert Review*
Consistency-Inconsistency error	Circulatory errors	√		
	Partition errors	√		
	Semantic inconsistency errors		√	
Conciseness-Redundancy error	Grammatical redundancy errors	√		
Completeness-Incompleteness error	Incomplete concept classification		√	√
Correctness-Class definition error	Identity-identify real world class definition		√	√
Clarity-Communication error	Class description communication error			√

- Clarity: focuses on how clear and understandable a knowledge representation is. The class description communication error was used to measure the clarity of both the ontologies.

Three ontology evaluation tools were used to evaluate the ontology. These are: Protégé automated ontology reasoners, competency questions, and expert review. The automated reasoners are add-on applications in the Protégé ontology editor to check consistency of the knowledge representation (Protégé, 2014). Three reasoners—FaCT++, Pallet, and RacerPro 2—were used to check consistency of both the Trans_Dom_Onto and TCA_Onto. Both ontologies were found to be fully consistent by all of the reasoners.

The competency questions were also used to check the consistency, completeness and correctness of both the ontologies. A competency question represents certain requirement that the ontology should be able to answer. A set of competency questions was formulated for each requirement identified for both the ontologies. The list of competency questions is long; a few are presented here as examples.

Some of the competency questions related to communication transaction-modality are as follows:

- Are transactions designed based on different design patterns?
- Are transactions defined based upon the exchange of the physical, financial, and information resources?
- Are transactions defined based on whether it is external or internal to the organization?
- Are transactions designed based on the response timings?

The competency questions related to domain transaction-modality includes:

- Are transactions designed and grouped according to bi-lateral and multi-lateral collaboration?
- Are transactions defined based on the sector or application area?
- Does the transaction design incorporate different modes of the project delivery as one of the governing factors in the design of transactions?

The competency questions for message modality include:

- Are messages defined based upon the function they perform in an information exchange scenario?
- Are messages classified based on whether they are formulated as verbal or written messages?
- Does a message design incorporate how information is to be represented in a message?

The competency questions devised to develop the TCA_Onto are as follows

- Does the TCA_Onto represent assets related to transportation systems?
- Does the TCA_Onto represent the tangible capital asset knowledge according to the notion of generalization-specialization of concepts?
- Is the tangible capital asset knowledge organized according to the notion of composition-aggregation of concepts?
- Does the TCA_Onto capture attributes of the tangible capital asset?
- Does the TCA_Onto incorporate different relationships between concepts?

For both the ontologies, each competency question was checked manually to measure the percentage compliance with each of the three errors: semantic inconsistency errors, incomplete concept classification errors, and identity errors. For a specific error, the percentage compliance is the sum of the competency questions in error divided by the total numbers of competency questions multiplied by 100. The compliance of each measure (in terms of the errors found in the concept description), was recorded as full-compliance (denoted as F, which means a competency question is error free for a specific measure), partial-compliance (denoted as P, which means a competency question is partially erroneous, i.e. in between the two extremes for a specific measure), and non-compliance (denoted as N, which means a competency questions is fully erroneous for a specific measure). The results of the competency question based verification for the Trans_Dom_Onto and the TCA_Onto is presented in Table 8-2. The results of the competency questions based verification of the Trans_Dom_Onto, and TCA_Onto indicate a satisfactory performance.

Both the ontologies were validated through three domain experts using a structured interview approach. Each of them had more than 15 years of experience in different civil engineering fields. They were extremely familiar with the transportation sector and moderately familiar with the water, wastewater and solid waste management sector. In addition, they were moderately familiar with information modeling and the process of communication formalization. A structured questionnaire was presented to the respondents wherein questions were organized according to three assessment criteria: clarity, completeness and correctness. For each question, a multi-sheet table was developed to reflect various concepts in rows and respondents' responses in the columns. The respondents were asked to rate a given concept on a scale of 1 (strongly disagree) to 5 (strongly agree) in each of the three assessment criteria. The responses were recorded for

Table 8-2. Competency questions based verification results

	Measures								
	Semantic Inconsistency Errors			*Incomplete Concept Classification Errors*			*Identity Errors*		
	Compliance with Measures in Percentage (%)								
Name of Ontology	*Full*	*Partial*	*None*	*Full*	*Partial*	*None*	*Full*	*Partial*	*None*
Transaction Domain Ontology	100			79	21		79	19	2
Tangible Capital Asset Ontology	100			80	20		80	20	

each concept and an average score was calculated. The average score ranged from 4 (agree) to 5 (strongly agree), which indicates that all the respondents were in universal agreement on the clarity, completeness and correctness of the knowledge represented in both ontologies. Moreover, an overall ontology validation assessment was conducted. The overall average rating of both the ontologies ranged from 4.67 to 5 on a scale of 5, showing that the results are satisfactory and the respondents were in full agreement on the clarity, completeness and correctness of the knowledge represented in the ontologies.

CONCLUSION

Currently, municipal organizations find it difficult to exchange information efficiently with other infrastructure agencies due to the heterogeneity of underlying data formats and manual communications. The growing trend is to transform these manual and ad hoc communications to a more formalized computer-to-computer communication. The issue is how to define these communications. To address the issue, an eight-step ontology-supported transaction formalism protocol was developed that the transaction development-personnel can use to formalize transactions for computer-based exchange of information. The development of the proposed protocol is beyond the scope of this chapter.

For computers to talk to each other to achieve message-based interoperability between the information systems of the infrastructure organizations, communications should be defined in a neutral format—the ontology. To support the design, management and implementation of transactions in the domain of infrastructure management, two ontologies were developed: Transaction Domain Ontology and Tangible Capital Asset Ontology. This chapter describes the modality-based abstract-level taxonomies of transactions, messages, actors/actor roles, information, and transaction attributes as part of the Trans_Dom Onto development. These are the core concepts required to define transactions and message templates. The modality-based Tangible Capital Asset Kernel Ontology is also presented, which represents four modalities of the tangible capital assets in the domain of infrastructure that were further extended to develop detailed taxonomies of the tangible capital assets in the transportation, water, wastewater, and solid waste management sectors. In addition, a taxonomy of the tangible capital asset attributes is presented.

An area of application was identified in the domain of infrastructure management to show how these ontologies were used to design, manage and implement transactions. The AI&CAR/TCA Reporting was identified as a transaction that has the greatest potential for IT improvement. The message templates of the proposed AI&CAR/TCA Reporting transaction were formalized using the information represented in both the ontologies. View 1 of the message template is presented in this chapter, which represents the header information and payload information that were captured from the Trans_Dom_Onto and TCA_Onto respectively.

As part of the evaluation, both the ontologies were verified and validated using a criteria-based approach. The results of the ontology verification are satisfactory indicating that the knowledge representation in both the ontologies is correctly modeled. There are some limitations associated with the development and application of the ontologies and the protocol. The ontology related limitations are: (i) defining a set of bi-lateral and multi-lateral transactions under the collaboration-based transactions to represent a complete set of collaboration transactions in the domain of infrastructure management, and (ii) the knowledge represented in the ontologies is limited to facility, transportation, water, wastewater, and solid waste management infrastructure sectors that can be extended to other infrastructure sectors, e.g. electricity, gas, and telephone, if message templates are to be defined for these sectors. The transaction formalism protocol related constraints are; (i) the proposed protocol focuses on message-based interoperability of the information systems of the infrastructure organizations while doesn't support model-based exchange of information, and (ii) the protocol is currently used in a manual manner that can be incorporated into a more complete system information and workflow management system.

In the future, the Trans_Dom_Onto needs to be extended to incorporate a complete set of collaboration-based transactions in the domain of infrastructure management. The TCA_Onto is required to be extended to incorporate the tangible capital asstes in other infrastructure sectors (like; electricity, gas, and telephone) so as to support the design of message templates in these sectors. Also, the proposed protocol needs to be tested in various application domains or industries to validate its' generality. The transaction formalization cycle duration time needs to be examined to check effectiveness and efficiency of the protocol objectively.

ACKNOWLEDGEMENT

The authors gratefully acknowledge the Canadian Natural Sciences and Engineering Research Council for their support of this research. In addition, this work was conducted using the Protégé resource, which is supported by grant LM007885 from the United States National Library of Medicine.

References

Allen, G. N. and March, S. T. (2006). "The effects of state-based and event-based data representation on user performance in query formulation Tasks." *MIS Quarterly*, 30(2), 269–290.

El-Gohary, N. (2008). "Semantic process modeling and integration for collaborative construction and infrastructure development." Ph. D. thesis, Dept. of Civil Eng., University of Toronto, Canada.

Fernández López, M., Gómez-Pérez, A., and Juristo, N. (1997). "METHONTOLOGY: From ontological art to ontological engineering." *Spring Symp. on Ontological Eng. of AAAI*, Stanford University, California, USA, 33–40.

Felio, G. (2012). "Canadian infrastructure report card." Municipal Roads and Water Systems, Vol. 1.

Gómez-Pérez, A. (1996). "Towards a framework to verify knowledge sharing technology." *Expert Systems with Applications*, 11(4), 519–529.

Gómez-Pérez, A. (2001). "Evaluation of ontologies." *International Journal of Intelligent Systems*, 16, 391–409.

Gómez-Pérez, A., Lopez, F. P., and Corcho, O. (2005). "Ontological engineering: with examples from the areas of knowledge management, e-commerce and the semantic web." *Advanced Information and Knowledge Processing*, Springer-Verlag, London.

Gruber, T. R. (1995). "Towards principles for the design of ontologies used for knowledge sharing." *International Journal Human-Computer Studies*, 43, 907–928.

Gruninger, M. and Fox, M. S. (1995). "Methodology for the design and evaluation of ontologies." *Workshop on Basic Ontological Issues in Knowledge Sharing*, IJCAI-95, Montreal, Canada.

Guarino, N. (1998). "Some ontological principles for designing upper level lexical resources." *In Proceedings of the 1st International Conference on Lexical Resources and Evaluation.*

ISO. (2006). "Information technology-business agreement semantic descriptive techniques." *International Organization for Standardization*, Part 4: open-edi Business Transaction Ontology, (ISO/IEC 15944-4).

McCarthy, W. E. (1982). "The REA accounting model: a generalized framework for accounting systems in a shared data environment." *The Accounting Review*, 554–578.

Noy, N. F. and McGuinness, D. L. (2001). "Ontology development 101: a guide to creating your first ontology." Technical Report, KSL-01-05, Knowledge Systems, AI Laboratory, Stanford University, USA.

OCCS. (2006). "A strategy for classifying the built environment." Table 36, Information, Edition 1.0. *OmniClass Classification System*, ⟨http://www.omniclass.org/tables/OmniClass_36_2006-03-28.pdf⟩ (May 29, 2014).

Osman, H. M. (2007). "A knowledge-enabled system for routing urban utility infrastructure." Ph. D. thesis, Departransaction mapent of Civil Engineering, University of Toronto, Canada.

Protégé. (2014). "Protégé is a national resource for biomedical ontologies and knowledge." developed by the *Stanford Center for Biomedical Informatics Research*, Stanford University School of Medicine, ⟨http://protege.stanford.edu⟩ (May 29, 2014).

PSAB. (2009). "Guide to accounting and reporting of tangible capital assets." Public Sector Accounting Board (PSAB), *Canadian Institute of Chartered Accountants (CICA)*, Toronto, Ontario, Canada.

RosettaNet. (2002). "RosettaNet implementation framework: core specification." ersion 02.00.01, ⟨http://www.rosettanet.org/dnn_rose/DocumentLibrary/ tabid/2979/DMXModule/624/Command/Core_ViewDetails/Default.aspx?EntryId=4730⟩ (May 29, 2014).

Uschold, M. and Gruninger, M. (1996). "Ontologies: principles, methods and applications." *Knowledge Engineering Review*, AIAI-TR-191, Vol. 11.

Yu, J., Thom, J. A., and Tam, A. (2007). "Ontology evaluation using wikipedia categories for browsing." *Proceeding of the Sixteenth ACM Conference on Information and Knowledge Management*, Lisbon, Portugal, 223–232.

Zeb, J. and Froese, T. (2011). "Design and management of transactions in the AEC/FM industry using an ontological approach." *3rd Int./9th Construction Specialty Conference*, CSCE, Ottawa. Canada, 1–10.

Zeb, J. and Froese, T. (2012). "Transaction ontology in the domain of infrastructure management." *Canadian Journal of Civil Engineering*, 39(9), 1–12.

Zeb, J., Froese, T., and Vanier, D. (2012), "State-of-the-art information technology use for municipal infrastructure management." *Journal of Information Technology in Construction*, Vol. 17, 179–193.

Zhang, J. and El-Diraby, T. E. (2009). "SSWP: A social semantic web portal for effective communication in construction." *Journal of Computers*, 4(4).

CHAPTER 9

A Framework for Regulatory Ontology Construction within AEC Domain

Lewis J. McGibbney*
Bimal Kumar†

Abstract: *Within the last decade or so, ontology engineering, in line with the proliferous expansion of the open/linked/semantic web of things, has infiltrated many facets of research and is now clearly ubiquitous even within the architecture, engineering and construction (AEC) domain. With specific focus on engineering of ontologies from a legislative perspective, current solutions leave many questions relating to best practice unanswered and provide minimal guidance to how, as engineers, we can accurately develop accurate, standards compliant ontological manifestations of legal documentation. We present our research in the form of a process framework for semi-automated ontology construction using building regulations as an example scenario. This work will appeal to engineers familiar with the preceding technicalities as well as newcomers to ontology engineering who are challenged with alternative manifestations of legal documents within the AEC domain. The objective therefore is to further enable improved uptake and cultural embracement of open government data using the platform provided by linked data paradigm for better communication of AEC regulations.*

---

*NASA Jet Propulsion Laboratory, California Institute of Technology, Pasadena, 4800 Oak Grove Drive, Pasadena, California 91109-8099, email: lewis.j.mcgibbney@jpl.nasa.gov

†School of Engineering and Built Environment, Glasgow Caledonian University, 70 Cowcaddens Road, Glasgow, Scotland, UK, G4OBA, email: b.kumar@gcu.ac.uk

INTRODUCTION

The domain of Legislative Informatics[1] has maintained growing levels of interest and popularity in recent years as a direct result of increased levels of maturity within highly influential partnering areas e.g. semantic web, ontology engineering, linked and open data paradigms to name a few. Additionally, pressure from globally distributed organisations concerned with the topics open (government) data, data transparency, interoperability, extensibility and reusability of data and standardisation has made clear argument for legislative data to be published and disseminated in more user-oriented representations. Although portals such as http://www.legislation.gov.uk aim to provide increased access to open legislation, many facets of parliamentary legislative output remain absent as open manifestations within such portals e.g. accompanying guidance and such. With specific focus on such legislative data within the AEC domain this work builds upon our previous research concerning the modelling of subsidiary legislation[2] (SL) such as building regulations/codes (typically published at national, state and local level) as open, linked data (McGibbney & Kumar, 2013). We target this facet of legislative output as such documents contain significant, extremely challenging domain specific ambiguities involving the inclusion of tables containing numerical data, fire, flame and combustion graphs, complex diagrams presenting accessibility requirements, mathematical formulae, functions and equations relating to energy performance etc. Additionally these legislative artefacts are typically authored and enforced via a localized governance model at state and/or local authority level. This of course presents significant problems for ontology engineers who wish to develop conceptual manifestations of the domain.

This work addresses the existing gap between the law, governance and regulation within AEC on one hand and the transition towards the increased use of emerging (and in some cases matured) technologies such as open/linked data and ontologies on the other. We distinguish the following facets within our roadmap covering the following key areas:

1. An overview of regulatory governance within AEC; with weighed focus on observable problems within current regulatory drafting practice. We bring Section 1 to a close by highlighting potential opportunities for improved user engagement with regulatory texts.

---

[1]It serves great purpose to define Legislative Informatics as the application of study involving structure and characteristics of information properties and associations within a legal environment and/or context. In this particular circumstance, we further define the term to cover the use of technology to drive such studies.

[2]We use the term *subsidiary legislation* synonymously to refer to *supplementary* or *delegated legislation* and further classify documents (which communicate such artifacts of the law) as the same. Their defined meaning within the context of this work covers the permission for governments to make changes to the law using powers conferred by an Act of a parliament (or similar legislative power). In the UK, delegated legislation of this nature is usually referred to as a statutory instrument.

2. Section 2 concentrates on legal document modelling as a maturing area with a strong provenance model. It expands to cover our attempts at advancing legal document and ontology modelling within the AEC domain. The practice of legal ontology engineering features heavily within this paper as we identify existing legislative vocabularies and assimilated resources for use within the framework presented in Section 3.
3. We consolidate the above by first proposing the principal argument that parliaments should further embrace the paradigm shift towards open data (as a data representation) and the web (as a tool) to better communicate law and governance to society. This section ties in with the platform of opportunities highlighted in the introduction concerning the development of better methods to improve engagement with regulatory information within AEC. We continue to present an ontology construction framework for AEC SL using building regulations as a driver. Whereas in our previous work (McGibbney & Kumar, 2013c) we focused on suitable data representation models, this work specifically targets the semi-automated engineering of local[3] legal ontologies for the AEC domain. Such ontologies can be embedded within structured documents as a means to complement existing legal document mark-up, or used independently in accordance with existing linked data tools and principles.
4. Section 4 provides some discussion and commentary on ontology engineering and its advances within AEC specifically.
5. We draw this work to a close with some concluding remarks on the suitability of our work in addressing engineering of legal ontologies, a brief discussion of the limitations of our research and suggestions for future work within legal ontology engineering within AEC.

Overview of AEC Regulatory Drafting Domain

The process of regulatory drafting (regardless of which legislative ecosystem one chooses to study) is a tremendously difficult process to model. In this context, one should accept our use of the term 'model', as indicative for a shift of thought and/or emphasis towards some activity which comprises the capturing of domain specific knowledge (which in this case is AEC legislation). This may further incorporate the procedures and compliance measures required to adhere to such legislation as passed by a parliament and typically incorporates the use of Information Technology (IT) in order to satisfy the afore-described. The primary aim of this work is to *open up* AEC legislative resources by leveraging existing web and markup technologies to expose and enable user oriented data to be used in more intelligent ways.

[3]In this context, the term *local* can be considered in its adjective form; meaning that generated ontologies are restricted to an immediate domain, that they are distinguishable between domains, and that they typically comprise an accurate image of some specific legislative characteristics.

Our driving justification stems from a feeling of frustration at the way these texts are drafted, commissioned and published, of how publishing organizations *expect* users of these documents to *just* use them correctly (which usually means comply with), taking little consideration for the compliance process involved at both professional and citizen level. We envisage a far greater understanding of legal texts could be achieved if they were communicated to their intended audience in a more appropriate manner.

There are of course practical issues associated with the above argument. It should be noted that the *de facto* situation described above exists predominantly because drafting workflows resulting in the dissemination of legislation have been developed and maintained over a lengthy duration. Additionally, they reflect the historically rendered requirements encountered by certain legislatures. Finally, it is of utmost importance that the legislative output of such workflows is *exactly* as passed within parliaments. It has become utterly clear to those working within this space that this remains a primary objective of legislative workflows even in fact if this compromises innovation through use of modern technology.

This being said, we argue that one possible root of the communication problem is that little or no existing literature concerning drafting of SL (with a focus on AEC) actually exists. Consequently, it is extremely difficult to quantify, in an attempt to improve, the eventual production of legislation as open, linked data (OLD). We substantiate on this as follows;

1. AEC is about designing and building not about documenting age old drafting processes or the events which lead to construction and engineering activities.
2. By the time literature written to deal with newer regulatory process actually makes it to press it is usually already outdated. This is certainly the case with dynamic topics such as energy performance, fire protection, the environment, etc.
3. Changes in the political landscape dramatically affect public sector government and local authority departments (usually tasked with the development and maintenance of regulations) meaning that the provision of supporting documentation is regularly relegated to a lesser importance over other duties.

To a large extent, the worry for organisations or persons concerned with the delegated tasks of drafting regulations for AEC is excellently described by Bett (Bett et al. 2003) "that innovative designs can conflict with prescriptive requirements . . . and are on the whole based on dated building technology and design solutions and do not anticipate trends in design or technology". It should be noted as a matter of importance that drafting workflows should in fact embrace the opposite, where technological trends should at a minimum be considered and that renewed effort should be made to embrace technological shifts within the development of regulations. This would better prepare all parties involved with the impending mammoth tasks of regulatory compliance defined by our legislation drafters. Our exploration of this topic now focuses on areas holding potential opportunities for improved document-user engagement within the regulatory drafting space.

Opportunities for Improved AEC Regulatory Document-User Engagement

As an act of principle, our development of the narrative contained within this section allocates sufficient consideration for realistic steps which can be taken to initiate improvement within the document-user space. By this, we mean that readers should consider the forthcoming suggestions as potential *low hanging fruit* which could/should be picked in an attempt to improve user engagement, interaction and inference of SL on a per document[4] basis.

A great deal of work has been done within the areas of Artificial Intelligence (AI) and Information Technology (IT) within Law. Casanovas & Sartor (2010) provide an array of examples including "Legal Information Retrieval, Electronic Data Discovery, Collaborative Tools (e.g. Online Dispute Resolution platforms), Metadata and XML Technologies (for Semantic Web Services), Technologies in Courtrooms and Judicial Offices (E-Court), Technologies for Governments and Administrators (E-Government), Legal Multimedia, Legal Electronic Institutions (Multi-Agent Systems and Artificial Societies), The Socio-legal Web (*Blawgs*) and the Web of Data (3.0)". It is however staggering to discover that within the regulatory drafting stage of legislation development, to a large extent much of this research effort is ignored. This seems rather counterproductive considering that regulation drafters struggle just as much with the governance and enforcement of regulations as professionals and citizens do with the compliance agenda. Consequently, it would not be improper if one were to question the fundamental efficiency and usefulness of the entire drafting practice currently in production within many of our public institutions. After all, we (within our individual societies) very rarely base the quality of either regulatory enforcement decisions or more generally the application of the law to concrete outcomes in a qualitative manner. It is in fact the case that quality benchmarking for regulatory governance is usually based on a quantitative basis favouring the speed of applications working through the system as opposed to the actual handing and decision making which such applications are subject to (McGibbney, 2013a).

It is therefore entirely appropriate for us to weight emphasis of discussion on the potential advantages which can be leveraged by embracing some of the above technologies aimed at providing a "basis for a strategy to counteract the silo effects which come into effect wherever data needs to be assembled from multiple different sources, by providing for this data a common mode of description, focusing on the common reality... data can become more easily integrated, more easily queried, and in principle more easily subjected to domain-transcendent modes of computational reasoning." (Casanovas & Sartor, 2010).

---

[4]For the sake of argument one may consider a *document* as a complete, individual regulatory artifact. In the case of regulations, it is commonplace for any one document to contain numerous sections and subsections respectively. It is also extremely common for such documents to reference, cite or leverage information directly from accompanying documents. We do not consider accompanying texts within our definition of individual documents.

With this in mind, we close this section by mentioning some of the lessons we learned and techniques we explored within our previous research whilst attempting to improve the flow of information between regulatory documents and users. Consider the following as an example scenario: If one were to pick any AEC regulatory document, one would immediately discover that the content covers every aspect of construction activities in explicit detail. It becomes clear that such documents are far too detailed for the majority of cases and contain too many options. What the average builder, professional or citizen really requires is sufficient information to enable them to comply with the regulations in the simplest and most cost-effective manner possible. If we consider another example involving building control officers, acting on behalf of local authorities, who are primarily concerned with whether a building complies with the requirements of the building regulations, we understand that in order to do this, they need to *see the calculations*[5]. The underlying question however relates to how one should succeed in accurately obtaining such calculations? Where can one find, for example, the exact policy and requirements for load bearing elements of a structure? The average U-value provisioning for glazing constructed as part of an external conservatory, classed as an external structure with a gross floor area of greater than 50 m^2?, etc. These are merely examples of typical barriers we expect to encounter regarding the manual interpretation of regulations within AEC. Some topics of our previous research within the area of metadata and legislative XML in this context, therefore, focussed on developing and implementing a methodology for publishing regulatory documents based on open datasets (McGibbney & Kumar, 2012), undertaking comparative studies to determine a suitable representation data model for UK building regulations (McGibbney & Kumar, 2013b) and implementing an intelligent authoring model for SL and regulatory instrument drafting within the UK construction and engineering industry (McGibbney & Kumar, 2013c). Additionally we have developed and implemented knowledge-directed information retrieval and management frameworks for dynamically changing SL such as energy performance building regulations (McGibbney & Kumar, 2011a) and web-based ontology-enhanced building regulation information retrieval applications (McGibbney & Kumar, 2011b).

Before we detail our contribution to AEC legal ontology engineering in Section 3, we first document important advances within the field of legal document modelling which have had detrimental impact on our own understanding of this area.

LEGAL DOCUMENT MODELLING AND THE AEC DOMAIN

As a discipline, legal document modelling has seen a hive of activity over the last decade or so. This builds from years of technological improvements driven by the explosion of information within many domains. Some examples include the

[5]One can consider this phrase sufficiently representative of the work which needs to be undertaken by AEC professionals to ensure that construction design and specification meets required regulatory levels of compliance.

proliferation of information across the Web, enhanced understanding within the field of mark-up languages, innovation relating to how systems can better manipulate and communicate with data, and of course the *coming of age* for mark-up languages as standardisation efforts acknowledge interoperability as key to a better, more user oriented web. Within the last 5 or so years, contributions from within the domains of law, governance and technology represent the fundamental cornerstone of legal document modelling, laying the ground for inter domain design theory and methodology. Some legal arguments cross the boundaries between legal document modelling and the cognitive science space with coverage of the difficulties associated with legal ontology building due to the tendency for successive stages of the process to become "blurred into the dynamics of the filed research and the knowledge acquisition process." (Breuker and Hoekstra 2010) Other work provides narrative on the inherent problems associated with legal documents generally and the textual content specifically. (Biasiotti and Tiscornia 2010) explore ontology engineering from a linguistic point of view elaborating on the difficulties proliferated throughout natural language specific to the legal domain. We consider such examples of critical importance to the development and direction of our own research as it becomes very clear that strong parallels exist between doctrinal *black letter* law and the very regulations we see within AEC.

Palmirani et al. 2010 provide a strong case (building on top of legal document mark-up) which furthers the argument that legal document modelling is multi-layered in nature involving "Text: part of the document officially approved by the authority with legal power; Structure of the text: part of the document that states an organization of the text; Metadata: any information that was not approved by the authority in the deliberative act. The metadata can involve document description metadata (e.g. keyword), workflow (e.g. procedural steps in the bill), lifecycle of the document (e.g. history of the document over the time), document identification metadata (e.g. URL, URI, URN and annexes); Ontology: any information about the reality in which the document act a role (e.g. for a judgment the juridical system concepts) or any concept called from the text that needs a modeling; Legal knowledge representation: the part of the interpretation and modeling of the meaning of the text under legal perspective." Figure 9-1 provides a conceptual overview of this description.

Although objective descriptions of this calibre are of utmost relevance to AEC regulatory document modelling, it is also important to observe additional peculiarities presented by AEC regulatory texts. Such documents contain significant, extremely challenging domain specific ambiguities involving the inclusion of tables containing numerical data, fire, flame and combustion graphs, complex diagrams presenting accessibility requirements, mathematical formulae, functions and equations relating to energy performance, etc. One would, therefore, expect additional investment (within both document modelling and ontology engineering) to be required at all levels of the above hierarchy before AEC regulations can be represented in a similar manner.

To date, it would appear that the cognitive thought processes behind current regulatory drafting practice, to a large extent, ignore advances such as

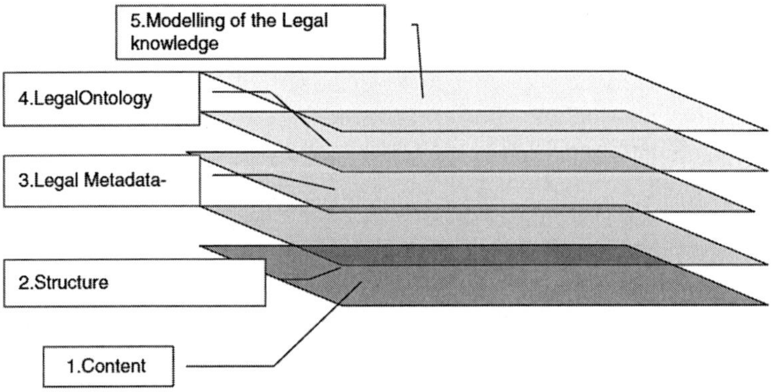

Figure 9-1. Layers of representation in legal document modelling (Palmirani et al. 2010)

the above[6] as there is little evidence of use within development or production environments. Because of this fact, we progress to explain justification behind our use and development of existing legislative vocabularies within our own work. The following section focuses on the methodologies, modelling trends and developments within legal ontology engineering which aim to provide increased use of legal knowledge. However, before doing so we conclude this section by reflecting on a statement provided by Breuker & Hoekstra (2010) which states that legal document and ontology modelling "... is a difficult problem, because ontology-building is a complex process, which can require ongoing knowledge acquisition efforts... lifecycles follow other patterns which may not be predicted in advance... Moreover, legal knowledge is dynamic". We strongly agree with this statement, as the problem description is wholly representative and also readily applicable to the AEC domain itself.

Existing Legislative Vocabularies

Recently published documents such as the Declaration of Parliamentary Openness (OpeningParliament.org, 2012) set clear guidelines for not only governments but also for authorities and organisations who wish to disseminate public data to promote, educate, make available and publish it in such a way as to make it easily accessible for citizens and business alike. The problem areas highlighted earlier concerning the development and subsequent publishing of AEC regulatory texts make strong argument of the failure of central governments to acknowledge the highly fragmented nature of the construction industry and the consequences this has on the authoring of such guidance. Although documents such as the

[6]It should be noted that by no means does the commentary and inclusion of literature (from within the domain of legal informatics) provided here aim to fully encapsulate the research contributions within the domain.

Declaration of Parliamentary Openness clearly acknowledge that the current situation needs to be improved, there is a huge amount of work to be done to accurately understand the best strategic methods required to achieve this utopian vision. The forthcoming sections therefore document existing legislative vocabularies (both mark-up models and ontology libraries) which can be utilised to aid the eventual manifestation of legal ontologies for AEC.

Legal Mark-up Languages

Our commentary on legal mark-up begins with the proposal for a Dutch Legal eXtensible Mark-up Language (XML)[7] Standard (Boer et al. 2002) which made progress on proposing an XML standard for legislative resources within the Netherlands. It is now clear that early work of this nature made significant contribution to the field of legal document modelling. We justify this by praising the logical aims proposed for standardizing legal concepts such as document filtering, presentation, document management, knowledge representation, search and information retrieval tasks etc. More importantly, however, we acknowledge the clear limitations brought about by individual mark-up languages (such as Dutch Legal XML) for country specific application. Independent models of this kind create barriers for interdisciplinary, cross domain interoperability and exchange of legal data. Although the model may work well within one given domain, it may not be sufficient and/or applicable for other use cases even within the same jurisdiction.

Similar initiatives around the same time (van Gog & van Engers, 2001) attempted to model legislation from a linguistic perspective using natural language processing techniques. The aim here was "not to perform this modelling in a batch fashion from legislation to final model, but interactively in dialogue with the knowledge engineer". Although one will note that the tangible outcomes of such studies were minimal, there is no doubt that they significantly advanced the domain of legal document modelling in general.

More recently a legal mark-up model was developed and is currently in use within the 'UK National Archives' www.legislation.gov.uk portal. The Crown Legislation Mark-up Language (CLML) (Sheridan, 2011) aims to encapsulate a governance model for the representation of all UK legislation regardless of its nature. In combination with other tools, CLML was developed to fit the stringent requirements of UK legislative data not only a means by which legislation can be accessed but more importantly as a machine processable concept representing the bones, the structure to which we attach the fleshy legislative content. During our work (McGibbney and Kumar, 2013b) we identified a number of issues where the CLML lacks the expressiveness required to model AEC regulations. Additionally it should be noted that the CLML (to an extent) is in conflict with the modern development ethos associated with legal mark-up, where integration of knowledge sources and interoperability are seen

[7]http://www.w3.org/TR/xml11/

as key to better communication and understanding of legislation. The UK specific nature of CLML means that its usefulness as a data representation format is restricted to UK jurisdiction. We therefore argue that this is a fundamental limitation of the CLML model.

Brining this section to a close we comment on two further models being used to model legislative artefacts and, therefore, of importance within the scope of this paper. Firstly, Metalex and Metadata Primer; "a jurisdiction-independent XML standard... that can be used for interchange, but also – maybe more importantly – as a platform for development of generic legal software." (Boer & van Engers, 2011), and Akoma Ntoso (AKN); an "XML schema for legal documents, a naming convention for the legal resource identification (URI[8]), and legislative drafting guidelines for leading the draftsmen to produce well-structured legislation documents and in the meantime suggesting best practices of quality assurance." (Palmirani & Vitali, 2011). In implementing metadata primers such as Metalex, regulatory document drafters are able to significantly reduce legislative information communication and interoperability across judicial and geographical boundaries. Similarly, through the use of AKN, drafters of regulatory documents can use a standard practice and language to produce and disseminate legislation. We provide detailed investigation and analysis of AKN as a suitable document representation model for building regulations in our previous research (McGibbney & Kumar, 2013b). Before we progress to detail legal ontology vocabularies, we conclude our discussion on legal mark-up stating that AKN also maintains and exposes tracking and change management for legal resources. As regulatory documentation is dynamic in nature, this aspect of the AKN model is utilised within our ontology generation methodology presented later.

Legal Ontologies

Within the confines of this section we intend to build on the definition of "ontologies as a type of knowledge representation... and formalization of legal knowledge" (Casellas, 2011). We specifically focus on forms of representation and formalization of legal knowledge and issues related to knowledge acquisition. The main objective of this section is to highlight how the use of legal ontology can be used for improved knowledge extraction and modeling methodologies, and within tools for ontology construction and ontology evaluation.

We, therefore, again focus observations towards early European research contributions such the CLIME Ontology (Boer et al. 2001) which proposed "a large-scale ontology... for the purpose of a web-based legal advice system." Unfortunately, in this case the ontology engineering was structured with a narrow area of application in mind (maritime information and legal exploration) and therefore only encompassed the following separate components "domain; A domain ontology of the design, construction, maintenance, repair,

[8]Uniform Resource Identifier. http://tools.ietf.org/html/rfc3986

operation, and inspection of ships.", and "norms; Mappings from rules in legal documents to deontic constraints that allow or disallows certain types of legal cases." Tangible outcomes of this research exposed noticeable failures concerning the limited expressiveness contained within the acquisition of legal knowledge for all uses of the contents of legal documents. As we now know, it is naive to assume that any one resource comprises all of the legal knowledge required to make consistent and justified decisions within some given domain. The same can be said for ontology engineering (and therefore eventual decision making) within AEC.

(Breuker et al. 2002) propose "various ontologies for the information management of documents of criminal trial hearings." within their attempts to improve legal information serving and knowledge management. One particular aspect of this branch of research (and associated work by the same authors) established "an 'upper' ontology—LRI-Core- that has the role of providing anchors and interpretation to the various legal domain ontologies." In principle LRI-Core would "provide a broad, rather than 'deep' conceptual structure for the typical *legal*, or legally relevant . . . " notions associated with legislative artefacts. LRI-Core contains several major principles, however the elements of relevance to our account herein include:

1. Objects and processes are the primary entities of the physical world. In objects energy and matter are distributed, so that objects participate in processes, while processes transfer or transform energy. Due to the complex nature of AEC regulations e.g. prescriptive or functional, we can make clear characterizations of concepts which have to be regulated.
2. Mental entities behave largely analogous to physical objects. In fact, one may argue that the mental world consists largely of metaphors of the physical world. A typical mental object is 'concept', and mental processes affect mental objects. Whether this fact is believed or not is an epistemological issue. Facts of belief and knowledge are mental objects consisting of concepts.
3. Time and space have also an ambiguous status. Related to occurrences, they provide positions of events and situations. However, as physical entities they provide the qualities of extension (size, life-cycle) of objects and processes (field, duration). Regulatory texts can for example have ambiguous status as they are dynamic artifacts which may have many non-static sections all in enforcement during any one particular moment in time.

Over time the concepts from LRI-Core were more clearly understood and accepted within the domain of legal ontology engineering, this resulted in convergence of several initiatives (including LRI-Core) into the Legal Knowledge Interchange Format (LKIF)-Core principle ontology for the legal domain. (Hoekstra et al. 2009) explain that "LKIF builds on a combination of OWL DL[9]

[9]http://www.w3.org/TR/owl-guide/

and a rule formalism, e.g. RIF-BLD[10], offering a classical hybrid solution." The fundamental aim of LKIF-Core is that it "... is intended to cover all legal domains, an LKIF ontology should focus on concepts that are typical for the whole range of legal domains, i.e. they should be sufficiently general and abstract." LKIF-Core provides support as it can be deployed as a resource for special, legal *inference*. Additionally certain facets of the ontology are designed to facilitate the knowledge acquisition process[11]. Finally LKIF-Core provides a standard (logically representative) terminological framework designed for the legislative domain which promotes and facilitates standard exchange of legal knowledge across multiple (possibly heterogeneous) knowledge bases.

During our use of the LKIF-Core ontology within mark-up and manifestation of AEC regulatory texts, we have learned that the ontology is designed (and intended to be used) in a clustered fashion, where a user can *pick and choose* what they require. Within the selection process, an engineer has the option to choose abstract, basic and/or legal concepts respectively then apply them within a modification or rule based framework. It should be noted that we have restricted the description to be vague intentionally as we do not intend to provide a completely comprehensive overview of LKIF-Core within this account[12]. Finally we should emphasize that similar to LRI-Core, there are many concepts within LKIF-Core which we have not used to date. The account of our usage within this document merely reflects our current experience.

We, therefore, bring our discussion on legal document and ontology modelling within AEC to a close by reflecting on existing research such as (Lui, & Lambrix, 2010) which highlights issues surrounding accuracy, validity and representativeness of ontologies. Such issues are of critical importance in ensuring consistency across/between mappings within the ontology engineering process and therefore deserve mention. In this context, ontologies within the domain of law and governance with specific focus on AEC are no exception. We consider accuracy (and subsequent quality) of target ontologies with crucial importance as ultimately their purpose and the processes within which they are to be used could be mission critical within a legal context.

---

[10]The Rule Interchange Format-Basic Logic Dialect (see http://www.w3.org/2005/rules/wg/bld/draft-2007-10-30). The first version of LKIF proposed that its most basic layer should consist of OWL DL and SWRL (http://www.w3.org/Submission/SWRL/), complying with standards of the Semantic Web. SWRL has since been replaced by a proposal for a rule interchange format, RIF that is intended to become the common denominator of several well defined dialects. It has intended that the underlying intentions of this proposal fit well with those of LKIF, whose interchange function can be supported through alignment with this initiative, however there are two disadvantages: (i) It will take some time before this proposal becomes a standard, and (ii) RIF compliant technology is most likely to cater for specific dialects which are not necessarily suitable as basis for LKIF.

[11]Here we refer to discrete physical elements or *things* from within the domain which will inevitably be modelled in some way within the ontology development.

[12]The LKIF-Core ontology is both large in scope (containing 15 modules) and also heavily annotated and justified from a legal and philosophical view point. This conversation is not wholly appropriate for application to the AEC domain.

At its core, the framework we provide in the next section addresses legal ontology engineering intended to improve semantic representation of legislative artefacts within the AEC domain. Currently, there is most certainly an existing void concerning the design and development of localised legal ontologies. Throughout the remainder of this work we address this area in an attempt to move towards improved regulatory decision making within the AEC domain.

AN ONTOLOGY CONSTRUCTION FRAMEWORK FOR AEC SL USING BUILDING REGULATIONS AS A CASE EXAMPLE

Focusing now on our main contribution to the advancement of ontology research for the AEC domain, we present *Leonto*[13]; a semi-automated ontology construction framework for AEC legal ontologies. We focus on documents within the field of SL to demonstrate our contribution in this area. Although in principle, the execution of jobs within the framework is a closed loop process e.g. there is no option to intervene in the process execution during runtime, we consider the generation of ontologies via this method a semi-automated process. This is due to the inherent complexities associated with legal documentation which additionally requires a degree of human determination regarding the quality and standard of resulting output ontology resources. It is appropriate to reflect upon an earlier statement which describes legal ontology modelling "as a difficult problem, because ontology building is a complex process, which can require on-going knowledge acquisition efforts... moreover, legal knowledge is dynamic." (Breuker and Hoekstra, 2010). Additionally, as explored throughout earlier sections, knowledge discovery is a non-deterministic process. An example of this is the eventual amalgamation of the LRI-Core and LKIF-Core ontologies, the realisation that ontologies and efforts such as CLIME, etc. were limited in scope to be of significant on-going contribution to the domain of law, governance and technology. Within our framework we've made best efforts to learn lessons from a wealth of previous research within the domain of legislative informatics generally and legal document modelling specifically. We therefore acknowledge that such frameworks should be monitored, maintained and operated in line with sustained understanding of advances in areas relating to law, AEC and technological innovation. This section therefore provides consolidation of various technologies covered in previous content, before describing the framework in detail.

Using the Linked Open Data Paradigm to Take Regulatory Workflows to the Web

Globally, social and political perceptions regarding how public data is obtained or produced, authored, amended and disseminated have changed beyond all formal

[13]https://github.com/lewismc/leonto

recognition. The fundamental driver behind this global paradigm shift is our necessity to increase the openness and transparency of our democratic states, whilst in the process embracing our digital world by linking past, present and future segments of our society. Many individuals within the political sciences argue that open data is the digital fuel of the 21st century. A huge part of this process involves making Government information more accessible to its users. With this in mind, our driving motivation is to enhance the construction and development of AEC legal ontologies. In our recent work (McGibbney & Kumar, 2013c) we propose a framework for intelligent authoring of SL and regulatory instrument drafting within AEC. The framework effectively embraces open data standards, using CLML, AKN, Metalex and Metadata Primer to suitably mark-up AEC regulations. Additionally we use embedded legal ontology structure and existing vocabularies such as LKIF-Core, Metalex, Web Ontology Language (OWL)[14], FRBR[15] axioms to further annotate AEC regulatory documents in an attempt to improve their usefulness across the web. We do this by first exposing, then enabling the connection and assimilation of related data that wasn't previously linked with the intention of using the Web to lower the barriers to linking data. Our justification for investing time and effort into the early stages of data modelling is simple; the more structure we can both syntactically and semantically embed within regulatory data, the fuller, more comprehensive, local ontologies we can construct. The next section explores our ontology construction framework entirely detailing our work within this area.

Leonto: A Framework for Semi-Automated Ontology Construction

The use of document pipelines is common place throughout the data acquisition, publication and dissemination space. In this case, our framework involves a series of data fetching, Media-type detection, validation, metadata extraction and finally serialization phases which are executed over AKN XML. Such operations are only made possible by the presence of fine grained structure within input data, hence our justification behind prerequisite investment within the document modeling phase. Our principal tool of choice for workflow management and task execution is Apache Any23™[16]; an open source Java™ library that parses and extracts structured data in Resource Description Format (RDF)[17] from a variety of Web documents.

The framework operates on the principle that fully annotated AKN XML is converted to a data stream which travels through specific phases of the pipeline. Each phase in the pipeline executes a different processing task over the data stream as it flows through the pipe. The output results in triples which represent implicit metadata which can be used to build an ontological manifestation of

[14]http://www.w3.org/TR/2012/REC-owl2-overview-20121211/

[15]http://www.vocab.org/frbr/core.html

[16]http://any23.apache.org

[17]http://www.w3.org/RDF/

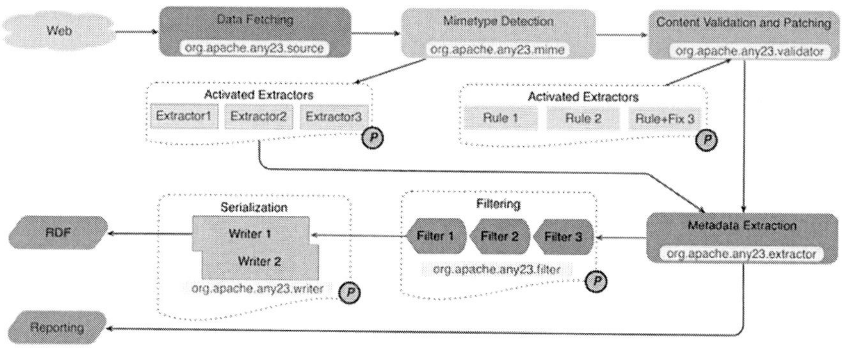

Figure 9-2. Apache Any23 Architectural Overview

the original AKN XML document. In order to achieve this goal we implemented a task specific extractor and parser module which extends common Any23 functionality. This involves extracting specific AKN metadata elements, their attributes and the corresponding data in order to achieve mappings between structured data and ontological artefacts. Extracting structured data from source resources involves leveraging existing legal ontologies mentioned above (LKIF-Core, FRBR, etc.) in combination with other RDF-based vocabularies such as Dublin Core[18], rdf-schema[19], etc. to produce fully localised document specific ontologies. An overview of a typical pipeline is shown in Figure 9-2.

A detailed description of Figure 9-2 is as follows[20]. We also provide commentary on our specific extensions to achieve parsing and extraction from AKN XML:

1. Content Sourcing and Fetching (org.apache.any23.source): We begin on the top left of Figure 9-1 where content is sourced and retrieved (fetched) either locally (File Transport Protocol) or from the Web (Hyper Text Transport Protocol). It is important to lay emphasis on the fact that Any23 acknowledges and makes best attempts to prepare for sourcing of raw data directly from the Web. In such instances we should consider that raw data of this nature will most likely contain incomplete, incorrect or inconsistent markup which will result in unpredictable results for us when generating ontologies. In our case, our source AKN XML data maintains a well-formed

[18]http://dublincore.org/

[19]http://www.w3.org/TR/rdf-schema/

[20]It should be noted that this account of the Any23 workflow highlights the core packages required for engineers to get up to speed with the Java library and adapt it to suit their own requirements. The account here does not aim to be a *one size fits all* solution. As is often the case with open source software there is a degree of investment required on the developer and user part. Additionally, the absence of content associated with the intermediate extractors is intentional as we only wish to discuss out own implementation here.

consistency of data meaning that levels of predictability (with regards to ontology generation) are improved considerably.

2. MIMEtype Detection (org.apache.any23.mime): In a sense, this comprises a very light weight data analysis (using Apache Tika™[21]; an open source Java™ parsing library), which evaluates the data type and makes an educated guess at which parser to use to read the data. In our case we implemented a custom package including a parser, extractor, parser report and exception handling classes designed to handle AKN XML. Our parser implementation utilizes a third party Java library[22] to achieve accurate parsing of the input AKN XML. We substantiate on our implementation in (4). The detection phase actually provides an indirect mechanism for measurement of data quality prior to subsequent meta-data extraction as automated guessing of MIMEType provides the opportunity for more than one type of extractor to be used in identifying structure embedded within the data stream. If the optional content validation and patching stage is not required, the pipeline typically skips to metadata extraction; detailed in (4). In some cases content validation and patching can be expensive due to the execution of rules over input data streams, (Mühleisen and Bizer, 2012) cover more on this topic and we briefly discuss this within our closing discussions.
3. Pre-determined rule-based content validation and patching (org.apache.any23.validator): As explained in 1 above, this stage is occasionally required due to issues such as bad structure, incomplete or incorrect mark-up, etc. typically present in a significant portion of data available on the Web. Resolving such discrepancies effectively improves the extraction of meta-data. To overcome such problems Any23 maintains a mechanism to detect such issues and in most cases to fix them. The detection and fixing is performed using an extensible collection of Rules. Such rules are written on a case-by-case basis usually dependent upon prior knowledge of the source input data. Currently data validation and patching is applied only on DOM[23] based documents, as AKN is XML-based it would be a very good application for our use case however it should be noted that we do not require content validation and patching as our source AKN XML is of a high data quality.
4. Metadata extraction (org.apache.any23.extractor): Metadata extraction comprises of data parsing, identifying and extracting certain aspects of the structure present within the AKN XML mark-up. Examples of such structure include the document title, its publication date, the previous amendment date, the document change revision number, etc. Within the context of SL however, additional metadata includes whether the document is part of a schedule or works, the country or county it was issued in, the jurisdiction

[21]http://tika.apache.org

[22]https://github.com/kohsah/akomantoso-lib

[23]http://www.w3.org/DOM/

under which it is relevant and legally binding, the language the document is written in, keywords associated with the document content, the organization who published the document, etc. With regards to explicit regulatory content e.g. the actual regulatory text as oppose to metadata, examples of structure include authoritative notes, hyperlinks, important contextual keyword definitions as they appear in the text, etc. The sole purpose of our parser and extractor implementation is for them to generate RDF statements representative of implicit (metadata) and explicit (formal text) semantic structure present within AEC regulations. We provide graphic examples of structured data extraction in the latter part of this section. The accuracy of these RDF statements is key to the usefulness of the ontology so it is key that as much work as possible (with regards to data modelling) is done upfront prior to data entering the pipeline in an attempt to lessen the burden on extractors. It should be noted that Any23 extractors work on the basis of being knowledgeable about certain vocabularies they may expect within the extraction process e.g. if an extractor is knowledgeable about the presence and/or occurrence of certain structure prior to runtime execution we greatly increase the effectiveness of extractors. We therefore invested a significant amount of time implementing extractor vocabularies for libraries such as LKIF-Core, FRBR, Metalex and AKN. Most of these implementations are now part of the Any23 codebase and available for the community to use.

5. Reporting: Optional reporting exposes behavioural characteristics of parsing and extraction implementations executed within the pipeline. Reporting content typically includes an account of the matching extractors used, the encoding used to parse the data stream, the detected MIMEType for any given document, whether content validation and patching was undertaken, number of triple relations actually constructed from the underlying structure and finally any issues which were encountered within the processing pipeline. In our own experience, we found extraction reports to be extremely valuable when assessing the quality aspect of our approach to ontology development as we were able to identify errors within the development and execution of our implementation.
6. Content filtering (org.apache.any23.filter): This complementary step enables us to root out and optionally remove *noise* from the extracted structure. An example of noise would relate to occurrences of nested structure within the data stream where we may want to limit how much of the structure we extract. Usually a determination of this granularity can only be made once a thorough understanding of the underlying data is achieved. We did not use content filtering during the development of Leonto, as again the underlying input data was clean and of a high quality.
7. Serialization (org.apache.any23.writer): The final stage of the pipeline typically involves writing the data stream out to one of the desired output formats. Any23 provides functionality to serialize data to JavaScript Object

```
<FRBRWork>
  <FRBRthis value="/uk/scotland/subsidiary/2010-07-11/4/main/schedule"/>
  <FRBRuri value="/uk/scotland/subsidiary/2010-07-11/4"/>
  <FRBRdate date="2010-07-11" name="Generation"/>
  <FRBRauthor href="#parliament" as="#author"/>
- <componentInfo>
    <componentData id="wmain" href="#emain" name="main" showAs="Main document"/>
    <componentData id="wschedule" href="#eschedule" name="schedule" showAs="Fire Subsidiary Document"/>
  </componentInfo>
  <FRBRcountry value="uk"/>
</FRBRWork>
```

Figure 9-3. Akoma Ntoso metadata mark-up snippet from Scottish Subsidiary Fire Regulation

```
FRBR:Work
  FRBR:this </uk/scotland/subsidiary/2010-07-11/4/main/schedule> ;
  FRBR:uri </uk/scotland/subsidiary/2010-07-11/4/> ;
  FRBR:date <2010-07-11> ;
  FRBR:author <#parliament> ;
  FRBR:componentData <Main document> ;
  FRBR:componentData <Fire Subsidiary Document> ;
  FRBR:country <uk> .
```

Figure 9-4. Snippet of generated Turtle displaying triple expressions of FRBR concepts

Notation (JSON)[24], NQuads[25], NTriples[26], RDF, RDF/XML[27], TriX[28] amd Turtle[29].

In Figures 9-3 and 9-4 we provide an example of triples generation from implicit regulatory metadata taken from the Section 4: Fire of the Scottish Technical Standards 2010 for Domestic Construction. The snippet of metadata in this example is located at the head of any AKN XML document and contains metadata which distinguishes the document itself (FRBRthis), the document URI (FRBRuri), the date the document was generated (FRBRdate) and so forth. Figure 9-4 presents sample triples output (ontology) serialized as Turtle for presentation purposes. The Turtle in Figure 9-4 contains 7 triples all of which are extracted from the implicit data contained within the FRBRWork node via the AkomaNtosoExtractor implementation mentioned previously. They should be interpreted as follows: 1; The document *work* has *this* expression */uk/scotland/subsidiary/2010-07-11/4/main/schedule*, 2; The document *work* has a *uri* value of */uk/scotland/subsidiary/2010-07-11/4/*, 3; The document *work* has a *date* value of *2010-07-11*, etc. It should be noted that whilst the output in Figure 9-4 may not be

[24]http://www.json.org/

[25]http://sw.deri.org/2008/07/n-quads/

[26]http://www.w3.org/TR/rdf-testcases/#ntriples

[27]http://www.w3.org/TR/rdf-syntax-grammar/

[28]http://www.w3.org/2004/03/trix/

[29]http://www.w3.org/TeamSubmission/turtle/

```
<section id="sec2-0-1">
  <num>2.0.1</num>
  <heading>Background</heading>
- <content>
  - <p>
    - <authorialNote id="atn1" placement="side">
      - <p>
          <a href="http://www.dontgive fireahome.co.uk">www.dontgive fireahome.co.uk</a>
        </p>
      </authorialNote>
      Life safety is the paramount objective of fire safety. Domestic buildings should be designed
      does occur, there are measures in place to restrict the growth of fire and smoke to enable t
      and effectively
```

Figure 9-5. Akoma Ntoso document text mark-up snippet from Scottish Subsidiary Fire Regulation

```
akn:section
  akn:num <2.0.1> ;
  akn:heading <Background> ;
  akn:content <Life safety is the paramount objective of
  akn:authorialNote <www.dontgivefireahome.co.uk> .
```

Figure 9-6. Snippet of generated Turtle displaying triple expressions of document text

immediately beneficial to general AEC activities, from a legislative perspective concerned with document provenance for example, the ability to express implicit semantic relationships in standardised vocabulary such as FRBR, etc. is extremely valuable. It enables us to query documents[30] (via their ontological manifestations) in ways not previously possible.

In the following, Figures 9-5 and 9-6 present examples triples extraction from explicit regulatory document text e.g. text that would be present in a hard copy of the regulatory document. Figure 9-5 shows a snippet of AKN XML markup again taken from Section 4: Fire of the Scottish Technical Standards 2010 for Domestic Constriction. The snippet represents Section 2.0.1: Background which contains an authorial note referencing a website as well as some paragraph textual content.

Based on the above AKN XML we can extract the triples as shown in Figure 9-6. Again these triples have been serialized into Turtle format for formatting purposes. They should be interpreted as follows: 1; A particular document *section* is *numbered* as *2.0.1*, 2; A particular document *section* has a *heading* value of *Background*, 3; A particular document *section* has *content* with value *Life safety is the paramount objective of...*[31], etc.

[30]It should be noted that a query model dedicated to this work is not provided here however the data representations included within this work can be queried by SPARQL the query language for RDF.

[31]The textual content as expressed in the triple output has been trimmed for formatting purposes.

DISCUSSION

In the previous section we provided a technical overview of the Any23 project, introduced Leonto; our customized extension framework specifically written to accommodate SL within the AEC domain, and presented examples of data extraction in order to produce ontological manifestations of regulatory documentation. This section now discusses aspects concerning current use of domain ontology[32] within AEC. Additionally we highlight some particular areas which, for us, have raised concern. This discussion results directly from our own experiences regarding improved understanding of ontology engineering as a discipline, the increased use of semantic web technologies generally and their application to AEC.

When one attempts to model some domain or facet of that domain there is significant tendency to design resources which are skewed towards some individual bias. In the case of ontology engineering this is dangerous. From an engineering perspective, the process of quality assessment and control becomes significantly more complex, as do other factors such as change tracking, provenance and validity/usefulness of particular resources (or indeed the entire resource itself) within the modelling domain as a whole. We consider the evaluation, testing and analysis of ontology resources as key to ensuring they accurately represent some concept from within AEC. To add to this, we are increasingly concerned with the development methodologies adopted by AEC ontology engineers when attempting to model some concept. Consider the example scenario where two engineers attempt to model one particular regulatory document. Both may first identify and then begin to formally represent trivial document characteristics such as the typical hierarchical structure present within documents of this nature. When, however, the development process extends to include areas such as the understanding and formal interpretation of natural and/or domain specific language, usage of complex legal axioms associated with many aspects of parliamentary output, etc. the potential for misinterpretation, bias, inconsistency and inaccuracy become significantly more prominent. Although we tend to consider ontology engineering as a practice which requires focus on interpretation (of information from within some given domain), technically and practically it requires more effort to ensure that information is not misinterpreted. In our research, we try to limit the degree of ad hoc human intervention and decision making during the ontology development process. This not only provides more consistent results but also enables us to standardise ontology development and tweak it to fit changing requirements from within the domain. From an ontology browsing perspective, the example provided in Figures 9-3 and 9-4 provides little benefit to those who wish to view comprehensive domain specific ontologies. Our justification behind the provision of abstract examples within these figures is simple. There is a wide and sweeping scope of change within regulatory and other legal documents across different legislatures. This is to say that successful usage of frameworks such as that proposed within

[32]As a natural successor to metadata and XML markup within this space.

paper is achieved on a case-by-case. The limited output included within Figure 9-4 represents only document level metadata as triples as it supports our argument backing the standardized and consistent development of ontological manifestations for AEC legal documentation.

We see many examples e.g. distributed systems, information retrieval applications, etc. using domain ontology in an attempt to improve communication of information. Whenever the question is posed regarding the implementation of the ontology design and development methodology, the quality of the underlying resource(s), how representative such resources actually are of the domain, etc. many researchers struggle to align their ontology engineering practice with these fundamental principles. This relates entirely to our example above, whereas in practice it is easy to observe that individual bias largely drives the ontology development process.

As we've mentioned before, legislative resources are dynamic in nature. This is a problem for persons within the legal domain and ontology engineers alike. Although legal mark-up languages such as AKN provide a change tracking mechanism for capturing such knowledge, very few people within the ontology engineering space acknowledge that if the domain, document and/or legislation which is represented by their underlying ontology changes, and that their system relies heavily upon this underlying ontology, then effectively their system is of lesser use than it previously was. Within the legal domain this can, on occasion, be mission critical. Say for example ontology resources were used within some information retrieval application to retrieve building regulations which were then used to make design and specific decisions. If the underlying ontology resource is outdated, inaccurate or relationships within the resource are inconsistent, users could easily make incorrect inference from the information resulting in inappropriate decision making.

Tracking changes within the legislative space is of extreme importance. Traditionally enforcement of regulatory resources is done via use of static hard-copies of documents. In some instances regulatory information is available online but usually the information is still represented in non-interactive formats such as PDF. One would have thought that with the availability of advanced inference and query technologies within the semantic technology ecosystem, we would have an appetite to infer, for example, if correct decisions were made on certain dates using certain building regulations. Only if data is structured, published and standardised methods are used to leverage the underlying data, will we be able to make such inferences possible. Unfortunately, currently this is not the case.

CONCLUSION

Within this work our intention was to address two ends of a spectrum spanning the practice of ontology engineering within AEC. The first topic concerned the existing gap between law, governance and regulations within AEC, the other covering the increased use of semantic web technologies, such as ontology. In doing this we first documented an overview of regulatory governance within AEC highlighting problems of interpretation which users commonly associate with SL such as building regulations. Section 2 delved into legal document and ontology modelling and the significant progress which has been made within these domains within the last decade or so. We present our framework for regulatory ontology construction within the AEC domain. We provide technological description of the framework and an example local building regulation as output. Finally, we concluded this contribution with a discussion on observation of areas within AEC ontology engineering which both concern us and provide areas for improvement within the space. The opinion and commentary we express within the work are the result of many years of research within the field of ontology engineering within AEC so we hope this is if use to newcomers within the domain and experienced engineers alike.

References

Berrueta, D., Labra, J. E., and Polo, L. (2006). "Searching over Public Administration Legal Documents Using Ontologies" *Proceedings of the 2006 conference on Knowledge-Based Software Engineering: Proceedings of the Seventh Joint Conference on Knowledge-Based Software Engineering*, pg 167–175, IOS Press Amsterdam, The Netherlands.

Bett, G., Hoehnke, F., and Robinson, J. (2003). *The Scottish Building Regulations; Explained and Illustrated*, 3rd edn. Blackwell Scientific. Oxford.

Biasiotti, M. A. and Tiscornia, D. (2010). *Legal Ontologies: The Legal Perspective, Law Governance and Technology Series*, Vol. 1. Springer.

Boer, A., Hoekstra, R., and Winkels, R. (2001). *The CLIME Ontology*, Springer Verlag, Berlin.

Boer, A., Hoekstra, R., Winkels, R., Engers, T. V., and Willaert, F. (2002). "Proposal for Dutch Legal XML Standard." *eGOV2002, State of the Art and Perspectives*. Springer Verlag, Berlin.

Breuker, J., Elhag, A., Petkov, E., and Winkels, R. (2002). "Ontologies for Legal Information Serving and Knowledge Management" in *T.J.M. Bench-Capon, A. Daskalopulu and R. G.F. Winkels (eds.), Legal Knowledge and Information Systems. Jurix 2002: The Fifteenth Annual Conference*, Amsterdam: IOS Press, 2002, pp. 73–82.

Breuker, J. and Hoekstra, R. (2010). *A Cognitive Science Perspective on Legal Ontologies, Law Governance and Technology Series*, Vol. 1. Springer.

Casanovas, P. and Sartor, S. (2010). "Foreword: What LGTS Intends to Be. Approaches to Legal Ontologies." *Law Governance and Technology Series*, Vol. 1. Springer.

Casellas, N. (2011). *Legal ontology engineering: methodologies, modelling trends, and the ontology of professional judicial knowledge*, Dordrecht: Springer.

Hoekstra, R., Breuker, J., Di Bello, M., and Boer, A. (2001). "LKIF Core: Principled Ontology Development for Legal Domain" *Frontiers in Artificial Intelligence and Applications. Law, Ontologies and the Semantic Web*. Vol. 188, pg. 21–52.

van der Hof, S. and Groothuis, M. Maria (2011). *Innovating government: normative, policy and technological dimensions of modern government*, Berlin: Springer.

Lui, Q. and Lambrix, P. 2010, "A System for Debugging Missing Is-a Structure in Networked Ontologies", *7th International Conference, DILS 2010, Gothenburg, Sweden*, P. Lambrix and G. Kemp, Springer, eds., August 2010, pp. 50.

McGibbney, L. J (2013a) "Better Regulation of Construction Activity through Data Modeling" PhD Thesis, School of Engineering and Environment, Glasgow Caledonian University, Scotland.

McGibbney, L. J. and Kumar, B. (2011a) "A Knowledge-directed Information Retrieval and Management Framework for Energy Performance Building Regulations" *Proceedings from the 2011 ASCE International Workshop on Computing in Civil Engineering: June 19-22, 2011*, Miami, Florida, US.

McGibbney, L. J. and Kumar, B. (2011b) "The WOMBRA Project: A Web-based Ontology-enhanced Multi-purpose Building Regulation Retrieval Application for Scottish Technical Standards" *Proceedings from the CIB W78-W102 2011 International Conference*: October 26–28, Sophia Antipolis, France.

McGibbney, L. J. and Kumar, B. (2012). "A Methodology for Publishing Regulatory Documents Based on Open Datasets" *Proceedings from the 7th International Conference on Innovation in Architecture, Engineering and Construction (AEC)*: August 15–17 2012, Sao Paulo, Brazil.

McGibbney, L. J. and Kumar, B. (2013b). "A Comparative Study to Determine a Suitable Representational Data Model for UK Building Regulations" *ITcon* Vol. 18, pg. 20–39, http://www.itcon.org/2013/2

McGibbney, L. J. and Kumar, B. (2013c). "An Intelligent Authoring Model for Subsidiary Legislation and Regulatory Instrument Drafting within Construction and Engineering Industry", to appear in Automation in Construction, Elsevier.

Morgan, B. and Yeung, K. (2007). *An Introduction to Law and Regulation: Text and Materials, Law in Context Series*, Cambridge University Press

Mühleisen, H. and Bizer, C. (2012). "Web Data Commons—Extracting Structured Data from Two Large Web Corpora" *In Proceedings of the WWW2012 Workshop on Linked Data on the Web (LDOW2012).*

OpeningParliament.org (2012). "Declaration on Parliamentary Openness" http://openingparliament.s3.amazonaws.com/docs/declaration/1.0/english.pdf.

Prins, C. and Eifert, M. M. (2002). *E-government and its implications for administrative law: regulatory initiatives in France, Germany, Norway and the United States*, The Hague: T. M.C. Asher Press.

Parlmirani, M., Cervone, L., and Vitali, F. (2010). "A Legal Document Ontology: The Missing Layer in Legal Document Modeling," *Approaches to Legal Ontologies, Law Governance and Technology Series*, Vol. 1. Springer.

Ruwanpura, J., Mohamed, Y., and Lee, S. (2010). "Construction Research Congress 2010: innovation for reshaping construction practice" *proceedings of the 2010 Construction Research Congress*, May 8–10, 2010, Banff, Alberta, Canada. Reston: American Society of Civil Engineers.

Sartor, G. (2011). *Approaches to legal ontologies: theories, domains, methodologies*, Dordrecht: Springer.

Sartor, G. (2011). *Legislative XML for the semantic web: principles, models, standards for document management*, Dordrecht: Springer.

Sheridan, P. (2010). "Legislation.gov.uk" http://blog.law.cornell.edu/voxpop/2010/08/15/legislationgovuk.

Van Engers, T., Boer, A., Breuker, J., Valente, A., and Winkels, R. (2008). "Ontologies in the Legal Domain. Digital Government" *Integrated Series in Information Systems.* Vol. 17, pg. 233–261

Van Engers, T. M., Kordelaar, P. J. M., Hartog, I. J. d., and Glassee, E. E. (2000). "POWER: programme for an ontology based working environment for modelling and use of regulations and legislation" *Proceedings of the 11th International Workshop on Database and Expert Systems Applications.* pg. 327–334.

Van Gog, R. and Van Engers, T. M. (2001). "Modelling Legislation Using Natural Language Processing" *Proceedings of the 2001 IEEE Systems, Man and Cybernetics Conference.*

CHAPTER 10

Taxonomy Development toward the Domain Ontology of Construction Contracts: A Case Study on AIA A201-2007

Jia Niu*
Raja R. A. Issa†
Ivan Mutis‡

Abstract: *In construction contractual management, sharing experts' domain knowledge through ontology is a good way to narrow the knowledge gap between the domain experts and the construction management team. However, little work has been done on ontology taxonomy development in this domain. Based on a literature review on sharing domain knowledge by using expert systems, taxonomy development methods and existing classifications in construction, this study proposes a synthesized methodology for taxonomy development in the domain of construction contract. The main parts of this methodology are building a conceptual model initiated from the definition of the contract, and the adoption of root classes from existing classifications, as well as the iterative development and competency questions approaches. In the case study section, using the research results from several pilot studies, the proposed methodology was applied to the AIA A201 General Conditions of the Contract for Construction (2007) document at the textual level to develop a taxonomy. The resulting taxonomy was used to determine the initial validity of the proposed methodology. The taxonomy development methodology and the developed taxonomy itself are both valuable contributions in the quest to further develop ontology-based applications for sharing domain knowledge about construction contracts. Furthermore, the potential of the taxonomy in ontology-based applications*

*Rinker School of Construction Management, University of Florida, PO Box 115703, Gainesville, FL, USA32611-5703; PH(352)328-7483; email: jniu@ufl.edu

†Rinker School of Construction Management, University of Florida, PO Box 115703, Gainesville, FL, USA32611-5703; PH(352)273-1152; email: raymond-issa@ufl.edu

‡Rinker School of Construction Management, University of Florida, PO Box 115703, Gainesville, FL, USA32611-5703; PH(352)273-1169; email: imutis@ufl.edu

for assisting construction contract and claim management is discussed. In the future when these applications are fully developed, construction project managers can use them to help in their performance of their contract management duties in order to minimize claims and enhance project success.

INTRODUCTION

Problem Statement

The development of more sophisticated buildings and building systems has led to more complex and consequently, more risky projects in the Architecture/ Engineering/Construction (AEC) industry. This trend has made construction project contract management much more difficult, and consequently, claims and legal issues have become more and more unpreventable due to the increasing complexity and uncertainty involved in construction projects (Hackett and Dancaster 2000). Further, for the project contractual parties, the performance of contract management duties and avoidance of contractual claims have an important impact on project success. Contract management requires domain experts with comprehensive contract knowledge and professional insight. However, due to the restrictions of the project jobsite, a knowledge gap exists between the domain experts and the construction project team. For example, in practice, the personnel given the role for managing claims are in most cases, because of the uncertainty of a claim arising, selected in an ad-hoc manner, as opposed to having dedicated personnel in that role similar to estimators, planners and accountants (Vidogah and Ndekugri 1998). Therefore, it is not unusual to find that in most cases, these claim clerks are lacking in claim-related professional knowledge and experience and that substantial errors and omissions often occur in the claim preparation process, which may lead to a loss of the claim (Cross and deBessonet 1985). The aforementioned substantial knowledge gap heavily impacts the performance of project contract management duties. In addition, the correctness of a contract management strategy closely affects the risk of successful claims being filed against a company as well as the opportunity of making a successful claim. For a project manager who is in charge of the contractual strategy and decision making, insight on the implications of the strategy and the accompanying decisions is very important. Hence, a project manager also needs a claim expert's insight and knowledge to support their contractual strategy and decision making process. Therefore, comprehensive and professional contractual knowledge and experience can be an effective source to provide useful support for contractual strategy and decision making.

Motivation

The combination of Semantic Web and ontology provides the fundamental theory and method for conceiving a potential solution to narrow the knowledge gap between the domain experts and the construction management team. However, little work has been done on this topic. To explore this research idea, representing the experts' knowledge through ontology was proposed and considered as an innovative way to solve this problem (Niu and Issa 2012). At the initial phase of building the ontology for the construction contractual domain, a solid taxonomy is necessary to classify and organize the related concepts in the contractual relationships between project parties. At this point, little work has been done on building the taxonomy in this particular domain.

According to the conceptualization process in the two prevailing ontology developing methodologies (Noy and McGuinness 2002; Gomez-Perez et al. 2004), the task following the building of a glossary of terms is building a concept taxonomy (often referred to as "define the classes and the class hierarchy"). This task is the most important one in the ontology development process, since taxonomy is a Knowledge Organization System (KOS) serving as the "backbone" of the domain knowledge for organizing concepts (Yu 2011).

Overview

To develop the taxonomy for the domain knowledge of construction contractual knowledge, this study proposes a synthesized methodology. This taxonomy development methodology starts from a conceptual model generalized from fundamental contract law principles; and then utilizes the common root classes to categorize the concepts that appear in the target contract documents. For this methodology the relationship between "classes" and "concepts" is that, "class" represents "class of concepts", and "root class" represents "class of major root concepts". In order to prevent the confusion due to the mixture of using "concepts" and "classes", the rest of this paper, unless otherwise explicitly indicated, will use "classes" instead of "classes of concepts." Based on a set of rationales, certain root classes used in popular top level taxonomies and/or classifications (e.g. IFC) are selected and adopted to initialize the development of this taxonomy. In order to determine the scope limitations of the taxonomy and to assure consistency of its terms, the following two approaches are also used: competency questioning (Grüninger and Fox, 1995) and iterative development (Gruber 1995a).

Finally, the validity of the proposed methodology is tested through a case study that applies it to the textual content of the clauses of the AIA A201 General Conditions of the Contract for Construction document (2007). As a result, a taxonomy was developed which was used to determine the validity of the proposed methodology. The taxonomy development method and the developed taxonomy itself are both valuable contributions in the quest to further develop ontology-based applications for sharing domain knowledge about construction contract. Moreover, the potential ontology-based applications can be derived from this taxonomy in construction contract and claim care introduced to demonstrate the value of this research result.

LITERATURE REVIEW

Representing Claim Knowledge in Expert Systems

The significant applicability of knowledge sharing in the legal area was addressed more than three decades ago. In 1980s, scholars noted that the law is a near-perfect application area for knowledge representation. Legal knowledge representation is needed in conceptual legal information retrieval systems and in legal reasoning systems (Cross and deBessonet 1985).

In the construction area, one of the traditional methods for sharing the domain knowledge about contractual issues is by using expert systems. Based on the theories about knowledge representation and logic reasoning, expert systems can represent and reuse domain knowledge to some extent. In Artificial Intelligence, an expert system is defined as a computer system that emulates the decision-making ability of a human expert (Jackson 1998). The first expert system were created in the 1970s and they proliferated in the 1980s (Leondes 2002).

In the area of construction contractual claims, with the development of expert systems in the 1980s, a large number of studies on the application of expert systems in construction contract and claim analysis were conducted from the mid1980s to the mid1990s. Figure 10-1 shows a timeline of the development of these expert systems that was derived from an extensive literature review.

Drawbacks

All of the expert systems previously discussed are ruled-based and make use of the logic reasoning relationship among concepts. However, this methodology

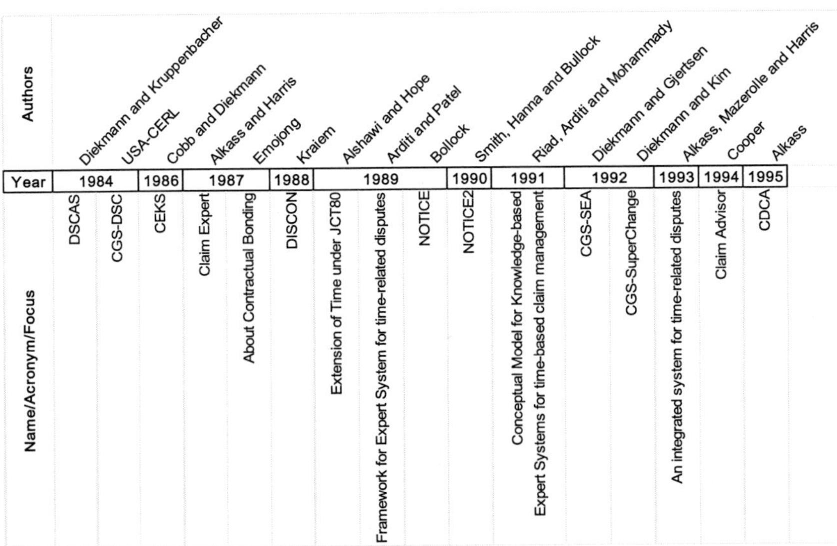

Figure 10-1. Timeline for expert systems development in construction claims

has shortcomings related to representing and sharing domain knowledge. The knowledge representation level achieved by expert systems is too shallow to reach the level of being semantically accessible and interpretable by machine. This shallowness mainly manifests itself in the inability of representing the concepts themselves. All of these expert systems do not represent knowledge from the base level of legal concepts that govern litigation outcomes (Mahfouz and Kandil 2012). To make up for this defect, the common practice used with expert systems is just throwing this problem back at the user by simply asking the user to figure out the concept interpretation and judgment. However, this practice makes it necessary for those using these so-called expert systems to be real experts, which defeats the purpose of using an expert system (Bubbers 1991).

As a matter of fact, the reason behind this problem is the absence of a global language which allows the knowledge bases to be shared by all systems. With a global language, the lack of implicit knowledge in one domain can be fulfilled by sharing the explicit knowledge in the other domains. The relationship between the Semantic Web and OWL (Web Ontology Language) in particular, to work in AI (expert systems) is somewhat parallel to the relationship between the Web and the hypertext community, it is based on some of the same motivations, but with a very different architecture that drastically changes the ways in which the technology can be deployed (W3C 2008). Berners-Lee et al. (2001) commented on the relation between expert systems' knowledge representation and ontology as "Knowledge representation (of expert systems) is clearly a good idea, and some very nice demonstrations exist, but it has not yet changed the world and that it contains the seeds of important applications, but to realize its full potential it must be linked into a single global system." In other words, knowledge representation in expert systems needs to be connected altogether and shared with all in order to achieve the breakthrough of next generation. Later, even though the agent-based approach was proposed in the negotiation of construction claims, the importance of domain ontology for knowledge was also emphasized (Ren et al. 2001). All of these problems show the necessity of using ontology and its ability to share domain knowledge in the area of construction contracts and claims.

Representing Domain Knowledge in a More Universal Way by Ontology

The term "ontology" comes from the field of philosophy that is concerned with the study of being or existence. In philosophy, one can talk about ontology as a theory of the nature of existence (e.g., Aristotle's ontology offers primitive categories, such as substance and quality, which were presumed to account for All That Is). In computer and information sciences, ontology is a technical term denoting an artifact that is designed for a purpose, which is to enable the modeling of knowledge about some domain, real or imagined (Gruber 1993).

Definition and Methodology of Ontology

Since the mid-1970s, researchers in the field of Artificial Intelligence have recognized that capturing knowledge is the key to building large and powerful systems. Researchers argued that they could create new ontologies as computational models that enable certain kinds of automated reasoning (Hayes 1985). In the 1980s, the Artificial Intelligence community began to use the term ontology to refer to both a theory of a modeled world and a component of knowledge systems. Some researchers, drawing inspiration from philosophical ontologies, viewed computational ontology as a kind of applied philosophy (Sowa 1984).

During 1990s, the definition of ontology had become more and more clear and was widely accepted by researchers. In the early 1990s, an effort to create interoperability standards identified a technology stack that called out the ontology layer as a standard component of knowledge systems (Neches et al. 1991). Associated with that effort, Gruber (1995) is credited with a deliberate definition of ontology as a technical term in computer science. Gruber originally defined the notion of ontology as an "explicit specification of a conceptualization". Later on, Borst (1997) defined ontology as a "formal specification of a shared conceptualization." This definition additionally required that the conceptualization should express a shared view between several parties, a consensus rather than an individual view. Also, such conceptualization should be expressed in a (formal) machine readable format. In 1998, these two definitions were merged by the definition of ontology as a "formal, explicit specification of a shared conceptualization (Studer et al. 1998)."

As far as the methodology for building ontology is concerned the first guidelines for developing ontologies, ENTERPRISE Ontology and TOVE (Toronto Virtual Enterprise), were proposed by Uschold (1995) and Grüninger and Fox (1995) respectively and refined later (Uschold 1995; Uschold and Grüninger 1996). According to Uschold's guidelines, the following process must be followed to develop an ontology: 1. Identify the purpose of the ontology, 2. Build it, 3. Evaluate it, and 4. Document it. As shown in Figure 10-2 they proposed capturing knowledge, coding it and integrating other ontologies inside the current ontology during the building process. However, these guidelines were only applicable to the specific case studies used and did not set a standard for all situations.

As a matter of fact, depending on the characteristics of different domain areas, the process of building ontologies can vary. Subsequently, additional theories about the methodology of ontology building were developed based on different

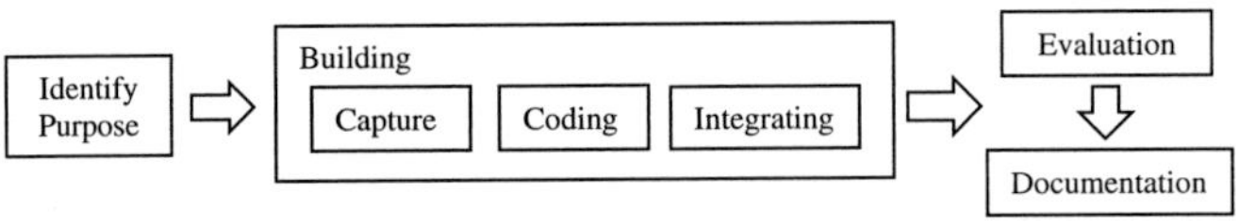

Figure 10-2. Main Processes of the Uschold's Method (Adapted from Gomez-Perez et al. 2004)

domains. The influential ones include: the KACTUS methodology (Bob et al. 1994; Guus et al. 1995) for reuse of knowledge about technical systems during their lifecycle in the manufacturing and engineering domains; the IDEF5 (Integrated DEFinition for Ontology Description Capture Method) method to develop and maintain usable, accurate, domain ontologies in software engineering. It is part of the IDEF family of modeling languages in the field of software engineering (Benjamin et al. 1994); METHONTOLOGY for building ontology for chemistry knowledge (Lopez et al. 1997; Lopez et al. 1999); guideline and methodology from natural language based SENSUS Ontology for machine translation (Swartout et al. 1997); On-To-Knowledge methodology for knowledge management systems (Staab et al. 2001); and the Noy and McGuinness' methodology (Noy and McGuinness 2002) consisting of seven steps.

Taxonomy in Ontology

However, in almost all of the influential ontology developing methodologies mentioned above, taxonomy (or class hierarchy) is an indispensable component for organizing concepts contained in a body of knowledge. Derived from its Greek stems (Lambe 2006), taxonomy is the science of classification. Originally, it referred only to the classifying of organisms. Now, it is often used in a more general setting, referring to the classification of things or concepts, as well the schemes underlying such a classification. In addition, taxonomy normally has some hierarchical relationships embedded in its classifications (Yu 2011).

Currently, taxonomy has many applications in different areas, like bioscience, librarianship, and information science etc. However, in the usage within Knowledge Management, taxonomies are considered narrower than ontologies since ontologies apply a larger variety of relation types (Suryanto and Compton 2000). Further, in Knowledge Management a taxonomy is a kind of controlled vocabulary known as Knowledge Organizing System (or Knowledge Organizing Scheme), which allows for the organization of concepts into concept schemes. In addition, it is also possible to indicate relationships between the terms contained in the scheme (Yu 2011). There are three basic characteristics of a taxonomy for knowledge management (Lambe 2006): 1. A taxonomy is a form of classification scheme; 2. Taxonomies are semantic; That is, they provide a controlled vocabulary to describe their knowledge and information assets, and express the relationships between terms in the taxonomy; 3. A taxonomy is a kind of knowledge map. A taxonomy should give its user an immediate grasp of the overall structure of the knowledge domain covered. The taxonomy should be comprehensive, predictable and easy to navigate.

In facilitating ontology-based applications, the advantages of taxonomy as a knowledge organizing scheme include: making searches more robust by related words matching instead of simple keywords matching; more intelligent browsing interfaces by following the hierarchy structure and by exploring broader/narrower terms; promoting reuse of knowledge and facilitating data interoperability through formally organizing domain knowledge (Yu 2011).

Methodologies for Building Taxonomy

Based on a literature review of taxonomy development in the engineering management area, it was determined that the methodology of content analysis is often used in finding a taxonomy from a large amount of textual materials (Chuan and John 2005; Goodman and Chinowsky 2000). Content analysis (or textual analysis) is a methodology in the social sciences for studying the content of communication. It gained popularity in the 1960s. Krippendorff (2004) defined content analysis as "a research technique for making replicable and valid inferences from texts (or other meaningful mater) to the contexts of their use." Typically, taxonomy studies using content analysis are mainly focused on determining the presence of certain words or concepts within texts or sets of texts, and then quantifying and analyzing the presence, meanings and relationships of such words and concepts to make inferences about the information in order to classify those words and concepts. However, this method has a large dependency on the text material selected which would bias the result. To minimize this bias, the application of content analysis needs a huge amount of literature sources (e.g. books, journals, documents, web pages etc.) to achieve adequate comprehensiveness.

To weaken the bias in the empirical approach discussed above, some other way of identifying taxonomy incorporating more theoretical concerns is needed. In the domain of knowledge management in construction, Lima et al. (2003) developed the Knowledge Management (KM) environment, e-CKMI, tailored for the Building and Construction (BC) sector in Europe. As a part of it, the e-COGNOS project addressed the need for developing domain taxonomy for construction concepts (El-Diraby et al., 2005). Besides the use of a search engine to find the frequency of concepts/terms, other tools and practices adopted include, briefly, using a process-oriented ontological model (equivalent to a conceptual model), which is an architecture of how the world (in a domain) behaves (or becomes) (El-Diraby, 2012), allowing utilization of already existing classification systems (BS6100, MasterFormat, and UniClass), and involvement of domain experts in intensive interviews, as well as the use of iterative development and competency questions. These tools and practices contribute to constructing taxonomy in a more theoretical sense, which makes the results more convincing and solid, compared to solely using content analysis.

Concepts in Existing Classifications for Construction

Considering the compatibility and reusability of the taxonomy to be developed, several influential conceptual models and taxonomies previously developed in the context of AEC processes were reviewed. The aim of this review was to identify some commonly used root concepts to initiate the conceptualization process for the desired taxonomy.

Conceptual Models for Construction

Luiten et al. (1993) developed the conceptual model of IRMA, an Information Reference Model for Architecture, Engineering and Construction. IRMA identifies

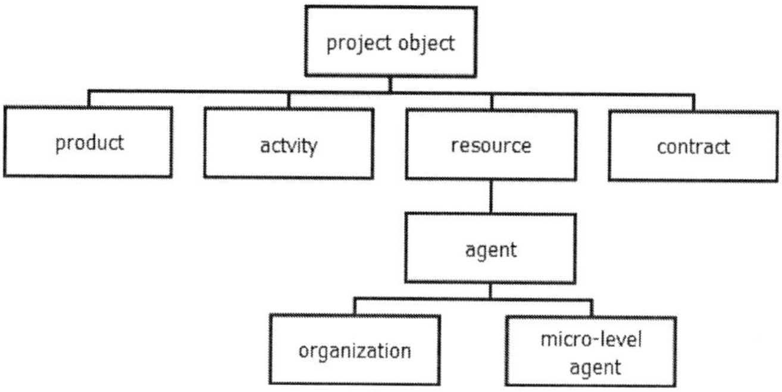

Figure 10-3. Inheritance relationship in IRMA (Adapted from Luiten et al. 1992)

and formalizes some of the key concepts of projects (such as physical components, activities, resources, etc.) and it shows the relationships between them, as shown in Figure 10-3. This model was developed by combining features from four previously developed conceptual models. The four models were: 1) Unified Approach Model as a generic model for modelling all kinds of construction information would facilitate the integration of different information technology applications from very diverse domains, such as CAD, project management, etc. (Björk 1992); 2) General Construction Object Model, GenCOM, developed to improve the integration of project management software using standard object-oriented models of construction projects (Froese 1992a, 1992b); 3) Building Project Model (BPM) developed by Luiten (1994) to provide a conceptual model that integrates product, activity, and resource information; and 4) the information/integration for construction, ICON, project (Aouad et al. 1994) for investigating the feasibility of establishing a framework for integrating information systems in the construction industry. IRMA was developed, based on these four model, as a new conceptual project model which identifies the central objects in AEC projects. Although IRMA was intended more as a reference and comparison tool than an end product, it has served as a useful vehicle for further conceptual development (Froese 1996).

Besides the conceptual models, the largest and perhaps the most significant product modeling standardization effort is the ISO Standard 10303, "Standard for The Exchange of Product Model Data", or STEP (ISO 1994; NPDERC 1995). It was jointly developed by numerous organizations to provide a computer-interpretable unambiguous method for exchanging product data to and from any system. Furthermore, based on the ISO 10303, IFC (Industry Foundation Classes), as the data model for describing building and construction industry data was developed. IFC defines an entity-relationship model consisting of several hundred entities organized into an object-based inheritance hierarchy. At the most abstract level, IFC divides all entities into rooted and non-rooted entities. Rooted entities derive from ifcRoot which is subdivided into three abstract concepts: object definitions, relationships and property sets. Among these three, ifcObjectDefinitions captures tangible object occurrence and types and are further subdivided into six fundamental concepts: ifcActor, ifcControl, ifcGroup, ifcProduct, ifcProcess,

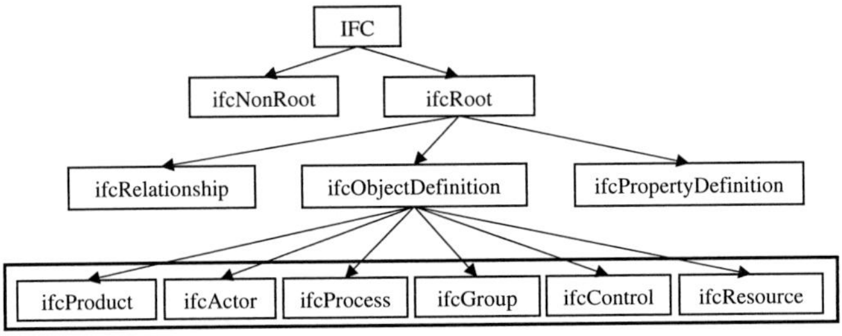

Figure 10-4. Excerpt of the IFC hierarchy regarding objects

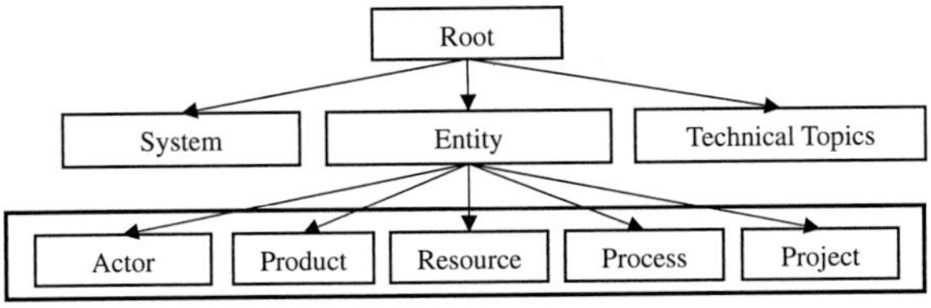

Figure 10-5. Excerpt of the e-COGNOS taxonomy (Adapted from El-Diraby et al. 2005)

and ifcResource (BuildingSMART 2006). The hierarchy is shown in Figure 10-4. These entity classes capture the core concepts in the construction process.

Domain Taxonomies

Additionally, with reference to the conceptual and data model mentioned above, a domain taxonomy for construction concepts was developed (El-Diraby et al. 2005). This taxonomy uses seven major domains to classify construction concepts: Process, Product, Project, Actor, Resource, Technical Topics, and Systems. The first five domains as shown in Figure 10-5 are very common in most taxonomies (Grüninger et al. 1997). They are clearly present in IFC and many of the construction classifications (OCG and the Building Catalogue, for example). Based on this taxonomy and other ones, El-Diraby developed Domain Ontology for Construction Knowledge (DOCK) (El-Diraby 2012) which continues to use that taxonomy and its major root concepts.

TAXONOMY DEVELOPMENT METHODOLOGY

The taxonomy development method varies in different domains. There is no "correct" way to model a subject domain - alternatives always exist. Therefore, considering the characteristics of the domain in this study, a synthesized methodology of taxonomy development for construction contractual knowledge is

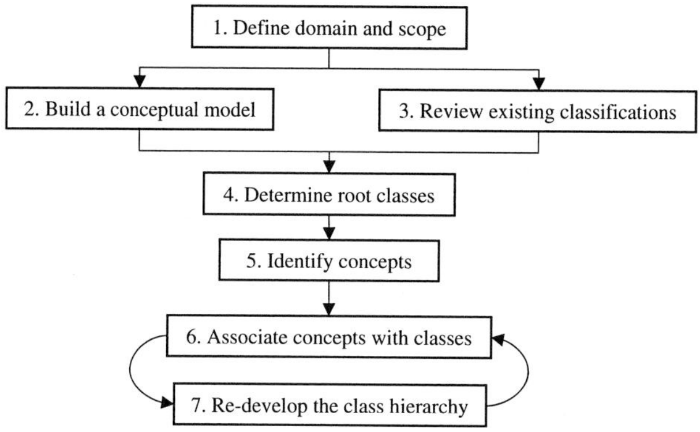

Figure 10-6. Framework of the Taxonomy Development Methodology

conceived. The framework of the taxonomy development methodology is shown in Figure 10-6.

1. **Define domain and scope.** To define the domain and scope covered by the taxonomy is the very first task. Clearly, in this study the domain is construction contract knowledge. It has to point out that the contractual knowledge and claim knowledge are closely related to each other, however, the knowledge scope of the taxonomy to be developed in this study only focuses on the construction contractual knowledge, not include knowledge body for construction claims.
2. **Build a conceptual model.** The conceptual model is a model to represent the core semantics of the subject domain. It is a model made of composition of concepts and can be used to help us know, understand, or simulate the subject domain. In particular, in this study, it is determined that the conceptual model is initiatively developed from the classic definition of "contract" and presented by textual description.
3. **Review existing classifications.** Considering what others have done is very important, since the existing works not only relate to the taxonomy's interoperability potential but also is a solid resource from where the candidate root classes of can be adopted. Since no existing classifications dedicated to the domain of construction contractual issue has been found, it is decided that to select some influential works in the context of AEC for review.
4. **Determine root classes.** Based on certain rationales for analyzing the candidate root classes from the existing classifications, the initial root classes for the taxonomy can be determined. Note that it has to simultaneously consider the availability of the candidates from the existing works as well as their applicability to the semantics of the conceptual model developed, as indicated in Figure 10-6.
5. **Identify concepts.** This step includes the discovering and extracting concepts/terms from various sources. In the case study section, the concepts are

identified from the text content of the AIA A201 Document. To control the level of details and the limits of the class hierarchy, competency questions method can be adopted in this step.

6. **Associate concepts with classes and 7. Re-develop the class hierarchy.** The interaction between these two steps is an iterative process until a consensus is reached. In particular, with more concepts associated with classes, it is very likely that the class hierarchy needs to be adjusted accordingly; the change of class hierarchy is bound to affect the decision on concepts' affiliations.

In this methodology, a synthesis of common mechanisms (PPC 2010) is applied. It mainly includes: Induction. Analysis of document for identifying concepts; Deduction. Adoption of the root classes from existing classifications; Inspiration. Viewpoints and judgments from the individual developer. In particular, the main steps or methods in the methodology framework are further elaborated in the following.

Step 2 - The Conceptual Model for the Taxonomy

This step is the one where the root concepts are initially introduced. The conceptual model has the most original and fundamental concepts for developing the taxonomy and is used as the basic reference for further decisions on adopting root classes. So, it is necessary to explain more on how to build the conceptual model in this study.

Since taxonomy is a concept scheme for organizing terms and concepts in a domain knowledge, the scheme needs to utilize some relationships among the concepts to organize them. Although the taxonomy for ontology development is not exactly the same as in designing classes and relations in object-oriented programming (Noy and McGuinness 2002), a taxonomy includes object-oriented features like encapsulation and inheritance (El-Diraby 2005). For example, the most common taxonomic relations, "*Subclass-Of*" comes from inheritance, and "*Is-a*" comes from polymorphism. Using these relationships, eventually, all the concepts can be categorized into a tree structure. The concept tree consists of several major root concepts (or root concept classes) that form its main branches. These root classes contain all the other specific classes as their sub-classes. The root classes themselves are at the top level of the whole taxonomy, and no classes can contain them. Further, a scheme is needed to organize these root classes. This scheme (also referred to as top level scheme) is able to describe the core semantics of what the target body of knowledge is. Particularly, in our case, the target body of knowledge is about the knowledge domain of construction contracts. Thus, the top level scheme focus on describing the essence of the construction contract and the top level scheme for organizing the root classes is defined as the conceptual model behind the taxonomy. To obtain the conceptual model in this study, the legal fundamentals of contract was studied.

The law applicable to construction projects falls into three major categories: contract, tort, and statutory/regulatory (Kelleher and Smith 2009). Since the scope of this study focuses on construction contractual issue, only contract law is

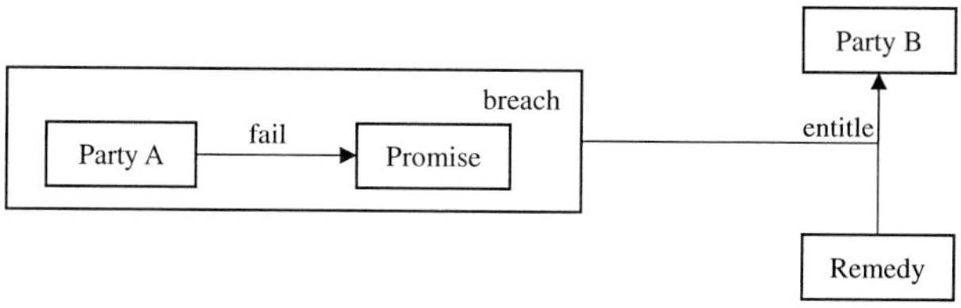

Figure 10-7. Illustration of the Conceptual Model for the Taxonomy

relevant. To initiate the building of the conceptual model, the definition of contract needs to be studied. Traditionally, the definition used for a contract comes from Restatement (Second) of Contracts (American Law Institute 1981) which defines a contract as "*a promise or set of promises, for the breach of which the law gives a remedy, or the performance of which the law in some way recognizes as a duty*". Thus, a contract is basically a set of promises made by one party to another party, and vice versa. Further, it also defines "breach of contract" as the result when one party fails in some respect to do what that party has agreed to do, without excuse or justification. In the context of construction, a breach of contract may be instantiated as, for example, a contractor's failure to complete the work on time, or failure to achieve the required performance of the work. Likewise, if an owner unjustifiably fails to make periodic contractual payments to the contractor as portions of the work are completed, that failure constitutes a breach of contact. Both "contract" and "breach of contract" serve as two fundamental concepts in construction contractual management. Therefore, based on their definitions, a naïve conceptual model for the taxonomy of construction contractual domain knowledge can be defined as "*When a project party unjustifiably fails to fulfill its contractual promises (obligations), a breach of contract occurs, which entitles the other project party to a corresponding remedy*". This conceptual model is illustrated by Figure 10-7.

Step 3 and 4 - Root Classes Adoption from Other Common Taxonomies

Existing classifications are used to provide candidates from which root classes can be adopted. Further, the rationale for picking root classes from the candidates is also necessary to be explicitly presented. By the end of this part, the root classes of for developing the taxonomy can be determined.

Besides the concern for the contract itself, since the context is the construction industry, the taxonomy development work took into consideration the scheme and content of some commonly existing taxonomies and classifications. Particularly, classification systems about the product model and/or process model (e.g. IFC and MasterFormat) provide us with existing external taxonomies to use when there is a need to refer to certain objects which belong to the product or process model. Thus, it is valuable to integrate existing classification systems by sharing certain root classes with them in order to make the desired taxonomy work together with them.

Table 10-1. Analysis of Existing Classifications

Name	Classes	Brief
IRMA	Project, Product, Activity, Resource, Agent, Contract	General reference model for building objects
IFC	ifcProduct, ifcActor, ifcProcess, ifcGroup, ifcControl, ifcResource	Date model for describing building and construction industry data
e-COGNOS	Project, Actor, Product, Process, Resource	Domain taxonomy for construction concepts

In the literature review section, the existing models and taxonomies have been reviewed are IRMA conceptual model, IFC data model and e-COGNOS taxonomy. In the context of this study, all of them can be deemed as the classification systems in the domain of construction, and be used as a reference for generating root classes of the taxonomy on construction contracts. However, to decide on the choice of classes from these classifications, an analysis has to be conducted on them and a set of rationales has to be set up for picking classes.

The high level classes from the three classifications are collected and listed in the Table 10-1. Note that in this table, IFC classes are the subclasses under the ifcObjectDefinition class and the e-COGNOS classes are from the Entity class.

The classes collected in this table provide a pool of candidates from where the root classes of the desired taxonomy are derived. Further, the set of rationales for deciding on the choices of classes is developed and explained as follows:

1. Choose the classes which are most in common among the classifications. Following this rule, it can be observed that the most commonly used classes are Project, Product, Resource, Actor, and Process.
2. The classes must be applicable in construction contractual domain and the irrelevant ones should be discarded. For example, the class of ifcGroup represents collections of objects for particular purpose such as electrical circuits. Obviously this is irrelevant to the contractual issue, so ifcGroup should not be picked.
3. If a class is equivalent to a potential concept involved in the conceptual model for the taxonomy to be developed, then that class should be chosen. To execute this rationale, it is necessary to understand the meaning of the candidate classes well and attempt to find their connection to concepts in the conceptual model. For instances, the Agent in IRMA is equivalent to Actor in IFC and e-COGNOS; since ifcControl represents rules controlling time, cost, or scope such as work orders, it can be understood as the contractual promises which regulate contract parties' behaviors. So does the Contract in IRMA. Thus, both ifcControl and Contract can be deemed as equivalent to the concept of "contractual promises" in the conceptual model.

Based on the rationales for picking classes from existing classifications for AEC area, it was decided that six root classes are selected and used for the taxonomy development. Specifically, the six root classes are: Project, Actor, Product, Process, Resource and Promise. According to the conceptual model defined above, these six root classes are included in the semantics of this model, whereas all the other classes that are not included are deemed as irrelevant and are ignored. Further, these six root classes are the most common and essential ones among all the classes in the reviewed models and taxonomies. Although a conceptual model can only consist of one layer of root classes and no sub-classes, it is assumed that in the desired taxonomy, each of the root classes will have different sub-classes as its descendants. Although these taxonomies using the six root classes are not specifically designed for the domain of construction contract, it is assumed that they could be used as a reliable basis for initializing the classification work and be able to be adapted and modified later on to meet the desired taxonomy's needs.

Competency Questions and Iterative Development

These two approaches are supposed to be applied in steps of 6 and 7, which are "Associate concepts with classes" and "Re-develop the class hierarchy", respectively. In particular, competency questions method is used for control the limit of the taxonomy, and iterative development is for achieving the consensus on the taxonomy development. Both of them are very important for the practicality and feasibility of developing taxonomy in practice. Thus, some more details of how to utilize the two methods in this study need to be elaborated.

Competency Questions

To control the scope of the taxonomy, the method of Competency Questions (Grüninger and Fox 1995) is adopted in this study. Note that although ontology and taxonomy are not the same and the competency questions method is originally for ontology development, it still can be borrowed for building a taxonomy. The reason is that the scope of the ontology is actually determined by the scope of its taxonomy.

Competency questions are a set of consistent questions that the taxonomy developer has to ask and adhere to in the development of each phase. These questions are designed for testing the taxonomy limits during the development process. In practice, the competency questions used to determine the taxonomy's limit include:

- Does the taxonomy contain enough information to answer these types of (competency) questions?
- Do the answers require a particular level of detail or representation of a particular area?

In this study, since the taxonomy is developed from the textual content of the contract clauses, a vast amount of details is easily encountered. Developers encountering too many details in the textual content, can have a lot of trouble extracting target concepts from the texts and can very likely get lost. To deal with this problem, the necessity of applying competency questions becomes more obvious. The competency questions can serve as the benchmark for testing a candidate concept's relevance to the target taxonomy and consequently control the taxonomy scope. Particularly, if it has to use a candidate concept for answering the competency questions, then that candidate concept is deemed as the target concept which should be encompassed under some class in the taxonomy; if not, then it is an irrelevant or over-detailed concept which should be discarded. By using this testing process, the developer can maintain control over the scope of the target taxonomy and prevent deviations from the original expected scope.

Iterative Development

Additionally, in the whole process of ontology development, the importance of the iterative development approach is proposed and emphasized by many ontologists (Gruber 1993; Gruber, 1995; Noy and McGuinness 2002; Yu 2011). As the skeleton of the ontology, the taxonomy development work also can borrow and use this iterative development approach.

As matter of fact, this iterative development has to be included in the taxonomy development process. The reason is that at the beginning of the first round of the classification, as the classes are short of instances to give the developer more sense and reference about the semantics, it is often hard to decide on the affiliation of a given concept to a certain class, e.g. a concept may seem to fit in more than one class, or a concept may look irrelevant to any class. As a result, after the first round, quite a few concepts could be left without any class affiliations. But, at this moment as more instances have been assigned to classes than at the beginning of first round, it become easier to reconsider the affiliation of yet unassigned concepts. This explains why it is necessary to adopt an iterative development strategy in this process, because iterative development allows for adjusting or updating the classification from an overall perspective which makes the taxonomy more consistent and integrated.

In summary, the main steps and approaches in the range of 1 to 4 are elaborated in this section. In the proposed taxonomy development methodology, they serve as the initialization process to prepare for further concept identification and classification work.

CASE STUDY

In this section, to illustrate steps of 5 to 7 in Figure 10-6, the proposed methodology is applied to develop a taxonomy from the text content of the AIA A201-2007 Document. In addition, the validity of the methodology is tested in this case study and the analysis of problems encountered in practice contribute to further modifying it.

Data Selection

To identify the taxonomy for the domain knowledge of construction contracts, the American Institute of Architects (AIA) Document A201-2007, General Conditions of the Contract for Construction (AIA 2007a) was selected as the knowledge source. The reasons for this decision include:

- AIA Documents as standard forms of contract are widely accepted and used in the construction industry. Thus, it could be deemed as a well shared domain knowledge source by the AEC community. A taxonomy developed based on this standard form can be beneficial to all its users;
- AIA documents have evolved over almost 125 years through numerous editions to become benchmark documents expressing the contractual relationships between construction parties (AIA 2013). During this long history, they have been through numerous revisions, which make them one of the most reliable knowledge sources in construction contractual domain;
- As integral part of the prime owner-contractor agreement, the general conditions of the A201 Document set forth the responsibilities of the owner, contractor and architect during construction (AIA 2007b). So, the contractual relationships regulated by the AIA A201-2007 Document falls within the scope of this research.

Data Preparation

The AIA A201 Document has a total of 330 terms listed in its Index section. These terms are called "Index" terms and provide a good initial source for taxonomy development. A pilot study was conducted by Niu and Issa (2013a) and a glossary of the "Index" terms was collected and reorganized. An excerpt of the glossary of reorganized "Index" terms is shown in Table 10-2.

Although the "Index" terms provide a good reference, a considerable number of them are not ready to be classified. For example, the interchangeable terms (e.g. "*Completion, Substantial*" and "*Substantial Completion*" are the same), are terms used as a reading guide without any real semantics (e.g. *Basic Definitions*, *Capitalization*, and *Interpretation*), and these terms actually contain a cluster of concepts and relations, like a "block" rather than one single term (e.g. like the form of "*Conditions related to* ..."). Additionally, some terms deemed as "attributes" instead of "concepts" (there are 37 terms deemed as attributes, e.g. *Representation*, *Effective date of insurance*) were also excluded. All of those unqualified terms were trimmed off from the initial 330 terms, which left 253 qualified terms ready for classification.

In another pilot study based on these qualified "Index" terms, Niu and Issa (2013b) developed an initial taxonomy with eight top-level root classes and the conceptual model was also modified from the naïve one created in the previous section. The resulting conceptual model with the root classes is:

- "Within the confines of the ***Environment***, a set of ***Actors*** are to produce ***Products*** with ***Resource*** consumption, in which their ***Behavior*** should follow

Table 10-2. Excerpt from Glossary of Terms

Name	Synonyms	Acronyms	Type	Description
Acceptance of Nonconforming Work	Nonconforming Work, Acceptance of	–	Concept	Owner accept Work that is not in accordance with the requirements of the Contract Documents instead of requiring its removal and correction
Acceptance of Work	–	–	Concept	Owner accept Work
Access to Work	–	–	Concept	Access to the Work in preparation and Progress wherever located
Accident Prevention	–	–	Concept	Duty of prevent accidents
Acts and Omissions	Action and inaction	–	Concept	The legal behavior of a certain subject
Addenda	–	–	Concept	A part of Contract Document, which is for additional material added at the end of it
Additional Cost, Claims for	Claims for Additional Cost	–	Concept	A kind of claim asking for cost
Additional Inspections and Testing	–	–	Concept	Inspections and testing occurred by the rejected Work
Additional Time, Claims for	Claims for Additional Time	–	Concept	A kind of claim asking for time
Administration of the Contract	–	–	Concept	A kind of obligation
...	...	...	...	...

certain ***Processes*** based on the ***Promises*** they made, otherwise a ***Remedy*** should be granted to make up for the extra ***Resource*** use if the Actors' ***Behavior*** is not excusable."

This conceptual model integrates the contract's definition as its framework and the inductive root classes as its elements. It provides a fundamental understanding of the construction contract's core semantics.

The "Index" terms are good enough for developing the top level of the taxonomy, however, they are not suitable for the further development of more specific classes, because most of them are considered as quite abstract and general. Especially since there exists a significant portion of "block" terms containing a cluster of concepts and relations. Additionally, among the top level classes of the taxonomy developed in the pilot study, some classes (e.g. *Promise, Behavior* and *Product*) contained a very large amount of sub-classes which needed further classification into a more detailed hierarchy. Therefore, to address that problem in this study, the test object was changed from the "Index" terms to the text of the AIA A201-2007 clauses.

Data Processing

To process the prepared data, steps 5 to 7 in Figure 10-6 are executed in this section. The rationale for how to identify concepts from the text content of clauses is presented. The points need to be kept in mind when associating concepts to classes and in re-developing the taxonomy hierarchy.

Completing this data processing work required several steps. First the meaning of all the contract clauses were read and well understood by the developers. Then, based on the modified conceptual model and the eight root classes from the pilot study (Niu and Issa 2013b), as well as the basic knowledge representation model of "`Subject-Predicate-Object`" from the ontology description language, the meaning of the text of each of the contractual clauses was analyzed and the corresponding concepts in it were identified.

It is necessary to emphasize that, the developer has to have a basic knowledge about construction contractual issues since he/she needs to learn and interpret the contract clauses well. The reason is that it is assumed that the decisions for concepts' identification and affiliation to a class are based on the developer's own judge. To assist in the correct interpretation, the developer can resort to certain references. For example, A201–2007 Commentary (AIA 2007b) provides explanations for many of the legal concepts of particular A201 provisions. Further, the language of the AIA A201 clauses is clear-cut and concise, and the articles and clauses are very well-organized. In particular, almost each one of the clauses is focused on one specific topic or concept. Thus, these wording and organization features associated with AIA A201 clauses provide the developer with many hints for identifying and classifying concepts from the textual content.

The procedure for how to analyze a clause and identify relevant concepts from it can be summarized as follows:

1. Read the textual content of the clause itself and the relevant reference to best interpret its semantics;
2. Simplify the sentence(s) into sense groups with reference to the triple model of "`Subject-Predicate-Object`". Then assign proper name to each group as the concept;
3. Find the concepts which have the logic relations of inclusion, parallel or inheritance with each other;
4. From the concepts in Step 3, identify the ones which are semantically related to the root classes or the concepts in the conceptual model.

To further elaborate this procedure, an example is presented in Figure 10-8. This example is from a sentence in the clause "1.1.1 THE CONTRACT DOCUMENTS" in AIA A201 Document. The sentence in the top box in Figure 10-8 is the original text. Based on its semantics, this sentence can be simplified into the sense groups as highlighted in the middle box in Figure 10-8. For instance, "The Contract Documents" is a sense group representing the concept of "Contract Documents", and "the Agreement between the Owner and Contractor (hereinafter the Agreement)" representing the concepts of "Agreement". Further, it can be observed that there is an inheritance relations among these concepts, as indicated by the structure in the bottom box in Figure 10-8. In particular, "Contract Documents" has the sub-classes of "Agreement", "Conditions", "Specifications", "Drawings", "Addenda" and "Other Documents". Also, "Contract Documents" is related to documenting the "promise", which is a root

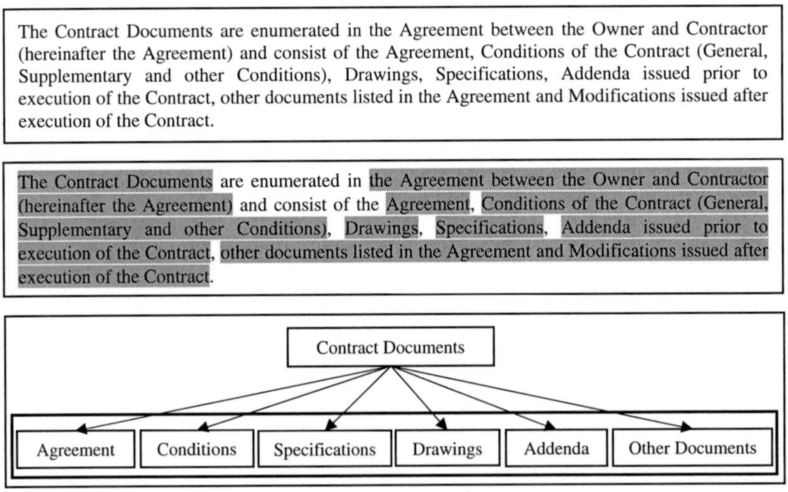

Figure 10-8. An Example for Procedure of Identifying Concepts from Text

class in the conceptual model. Thus, finally all of these concepts should be identified from the text.

Note that throughout this procedure, it is necessary to tell the difference between a "concept" and the "attributes/properties", and although the attributes/properties are also a crucial part of semantic modeling, here the focus is only on extracting the concepts from the text and ignoring the attributes, in order to avoid unnecessary confusion.

Once the concepts were identified from the text of the clause, each concept was put into the appropriate root class and the clause's code was recorded as the provenance information. The process was executed on one clause at a time, and then the same process was repeated on the subsequent clauses. With the progress of this development process, more concepts were accumulated under each root class. Using these accumulated concepts it became easier to identify the nuances among those concepts and to create appropriate intermediate classes to further classify those concepts. This process followed the middle-out approach for taxonomy development (Uschold and Grüninger 1996). Note that, for this case study, the scope of the work was limited to the first three Articles (General Provisions, Owner and Contractor) of the General Conditions of the AIA A201 Document for demonstration purposes. Since the contractual relationship between the owner and contractor is a core and typical part of the AIA A201 Document, it provided enough data for a good illustration of the application and validation of the proposed methodology.

Tool

The Ontology Annotation Tool (OAT) in GATE (General Architecture for Text Engineering) was found to be an efficient tool in completing the data processing task. GATE is a Java suite of tools originally developed by the University of Sheffield which is widely used for Natural Language Processing (NLP) tasks. OAT is a GATE plugin which enables a user to manually annotate a text with respect to one or more ontologies. The OAT tool supports annotation with information about the ontology classes, instances and properties. (Cunningham et al. 2011).

In this case study, since we are concerned with developing the taxonomy, the goal is to annotate a concept and link it to the class it belongs to. To be specific, both the AIA A201 Document PDF file and the initial taxonomy .owl file are first loaded into GATE Developer. Then the OAT interface is opened, the document and taxonomy are shown side-by-side (see Figure 10-9). In this user interface, a concept in a clause' text content can be easily annotated with a link to a class in the taxonomy, and the different links are color coded to distinguish them from each other.

With this feature, the above mentioned data processing procedure for this case study can be easily implemented. OAT's juxtaposition of both text and class hierarchy, and its enabling of links between concepts and classes productively facilitates to a large extent the building of the text-based taxonomy.

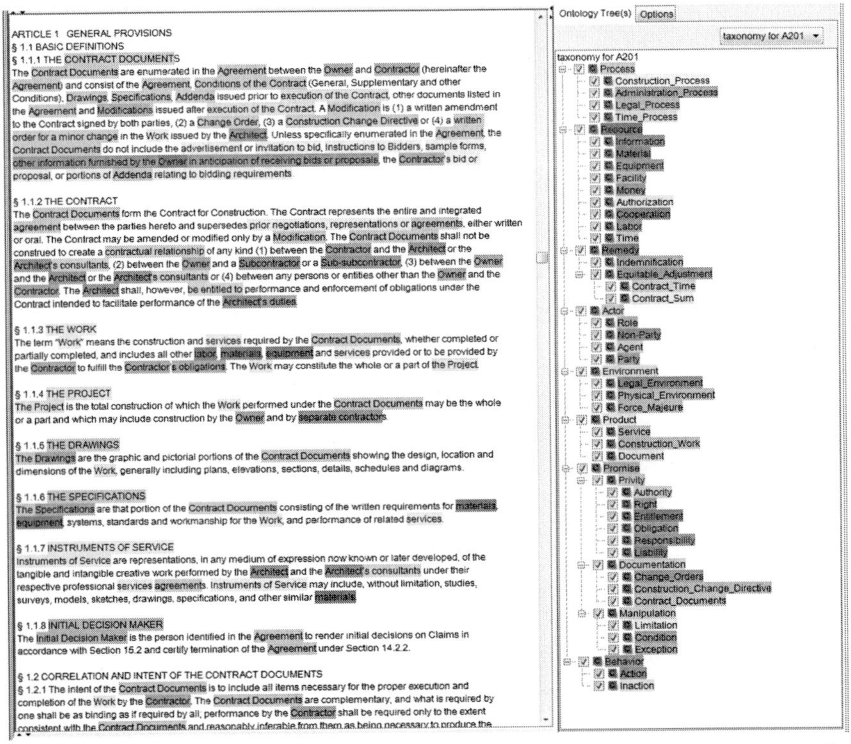

Figure 10-9. User interface of OAT in GATE for developing concept taxonomy for AIA A201

Results

The final results of this whole taxonomy development process are shown in Figure 10-10. The explanation of the root classes and the sub-classes are as follows:

- *Environment* which emphasizes the things out of the project *Actors*' control, and which could substantially affect the execution of the Contract. It has three major sub-classes:
  - Legal environment (e.g. applicable laws, statutes, ordinances, codes, rules and regulations, and lawful orders of public authorities),
  - Physical environment (e.g. site conditions, existing construction),
  - Force majeure (e.g. weather delay, labor dispute).
- *Actor* includes all the major players involved in the contract. This class includes the sub-classes:
  - Party (e.g. Contractor's superintendent, Architect's project representative), organization (e.g. contractors),
  - Agent (e.g. Owner's authorized representative, Contractor's authorized representative, superintendent),

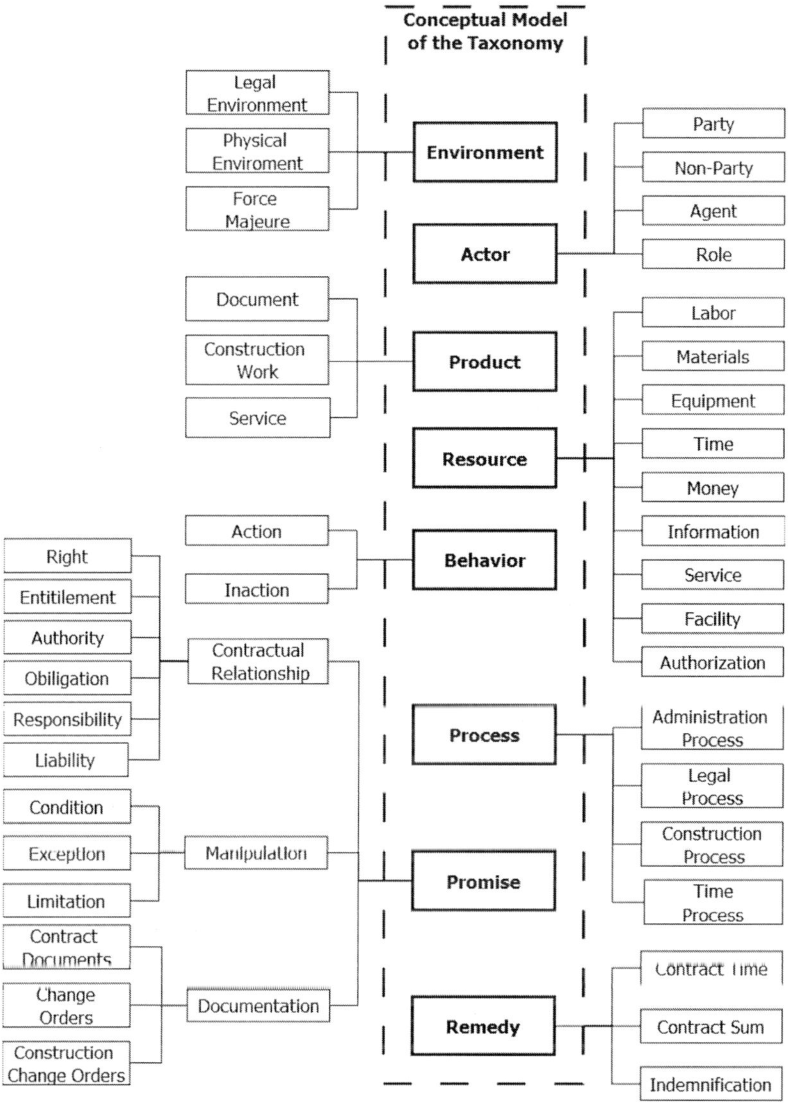

Figure 10-10. Higher Level Classes in the Taxonomy of AIA A201-2007

- Non-party (e.g. government agencies, municipality),
- Role (e.g. Initial decision maker, Surety).

- *Product* means all the needed items to be provided by a certain *Actor* as required by the contract. It may include tangible construction products (like a building, or a bridge), necessary construction activities to produce the construction products, and related service required by the contract. All of these products require a certain amount of *Resource* consumption. In short, the *Product* class has three sub-classes:

- Document (e.g. Instruments of service, submittals, schedules, evidence, record),
- Construction Work (e.g. the Project, the Work, construction or operations),
- Service (e.g. access to work, design service).

- *Resource* is an entity that has value, requires effort to obtain, and is something indispensable, i.e. without which certain needed work cannot be properly performed by an *Actor*. *Resource* is usually classified into five sub-classes. In addition to the usual five of *Labor*, *Materials*, *Equipment*, *Time* and *Money*, four more were added:
  - Labor (e.g. carpenters, masons),
  - Materials (e.g. plywood, cement, concrete),
  - Equipment (e.g. crane, scaffold),
  - Time (e.g. work day),
  - Money (e.g. dollars),
  - Information (e.g. information for preparing a mechanic's lien, information under the Owner's control and relevant to the Contractor's performance of the Work),
  - Service (e.g. service under the Owner's control and relevant to the Contractor's performance of the Work),
  - Facility (e.g. water, heat, utilities, transportation, and other facilities),
  - Authorization (e.g. approvals, easements, assessments and charges required for construction, use or occupancy of permanent structures or for permanent changes in existing facilities).
- *Behavior* represents the *actions* and/or *inactions* of the *Actors* which may lead to certain contractual consequences. Instances of *action* are commonly verbs in the clauses, which are abundant; whereas the number of instances of *inaction* is limited.
  - Action (e.g. authorize, execute, etc.),
  - Inaction (e.g. fail to, withhold, delay, ignore, etc.).
- *Process* contains certain procedures that should be followed by *Actors* for certain activities. It can be divided into four sub-classes:
  - Administration Process (e.g. Communications facilitating contract administration),
  - Legal Process (e.g. Meditation, Arbitration),
  - Construction Process (e.g. Construction means, methods, techniques, sequences),
  - Time process (e.g. bids are received, award of the Contract, commencement of the Work).

- *Promise* is about the contractual relationships among certain *Actors*. It is the core part of the whole taxonomy, so it has a more complicated structure than others. Specifically, *Promise* consists of three sub-classes, and each one can be further classified into several sub-sub-classes:
  - Contractual Relationship
    - Right (e.g. copyright, ownership, mechanic's lien right, right to stop the Work),
    - Entitlement (e.g. reply on the accuracy, increase in Contract Sum or extension of Contract Time, Change Order, reimbursement),
    - Authority (e.g. consent, enforce obligation against),
    - Obligation (e.g. report promptly, reasonable infer, secure and pay, indemnification),
    - Responsibility (e.g. jobsite safety, loss and damage, warranty, acts and omission of agent),
    - Liability (e.g. pay avoidable costs and damages, loss caused by patent or copyright infringement).
  - Manipulation
    - Exception (e.g. damage or defect caused by abuse, improper or insufficient maintenance),
    - Condition (e.g. failure of payment, material change in the Work),
    - Limitation (e.g. extent of indemnification).
  - Documentation (e.g. Contract Documents, Change Orders, Construction Change Directives).
- *Remedy* is deemed as a makeup for extra and excusable *Resource* consumption. Simply, it is mainly instantiated as:
  - Contract Time,
  - Contract Sum,
  - Indemnification.

DISCUSSION OF RESULTS

Improvement in Taxonomy

Compared with the taxonomy based on the "Index" terms of the AIA A201 Document (Niu and Issa 2013b), this taxonomy based on the text of the clauses keeps the eight root classes, but has a more specific and more detailed hierarchy of the sub-classes under each root classes. Especially, for some root classes which contain a large number of concepts, like *Promises*, substantial in-depth classification was done; and some more sub-classes were added into some root classes, like *Resource*, *Actors* and *Process*. Meanwhile, note that the conceptual model for the taxonomy, as marked by the dash box in Figure 10-10, is still the one developed in the pilot study without any changes, since the root classes are intact. Moreover, using the contractual clauses' textual content as the test object instead of the "Index" terms, avoids dealing with the problem of "block" terms, and allows the conceptualization work to be performed on a more detailed and finer level. This also gives the conceptualization a stronger capability for expressing the semantics from the very base level.

Significance of Competency Questions

During the completion of this case study, when conceptualizing the semantics in the textual content of the clauses, too many details and fineness was encountered making it difficult to determine which class needed to be added to the classification scheme. Particularly, sometimes it was difficult to identify and extract a concept from a group of sentences or a paragraph in a contractual clause, and thus to determine the concept classification scheme.

The method of competency questions was used to solve this problem. When the developer feels lost in facing too much detail, the competency questions work as a criteria for testing the boundary of the necessary level of detail that the expected ontology should reach. For instance, to trim off the irrelevant content, the question of "Is this concept closely related to the root classes in the conceptual model?" can be used as a benchmark. As a result, the AIA A201 contractual clauses like 1.3 Capitalization and 1.4 Interpretation can be easily identified as irrelevant ones and ignored with confidence. Also by applying the set of questions "Is this concept more like a class or a specific instance? Is it meaningful to let it have any further subclasses?" the level of detail can be controlled. For example, when considering the AIA A201 2.1.2 clause about the Owner's furnishing of information for the Contractor's mechanic's lien, this kind of information can be considered as a resource in the conceptual model, but the specific parts of the information were taken as instances and were not worthy of being assigned further subclasses. Thus, only the concept of "information for mechanic's lien" was selected for inclusion in the taxonomy as a subclass of Resource, but its specific parts were excluded.

The preceding is an example is of the process for testing and controlling the boundary of the detailed level of the taxonomy by using competency questions. The details beyond the boundary determined by the competency questions are considered as surplus and should be just ignored in order to reduce confusion. So, the use of competency questions is emphasized throughout the case study.

Compatibility with Ontology Description Language

In completing the case study (the process described under the Data Processing section), it was found that the structure of some contractual clauses can be very complicated, and some concepts are difficult to be defined and need verbose descriptions to express them clearly in English. However, the semantic modeling primitives provided by the ontology description language of RDF (Resource Description Framework) (W3C 2004), RDFS (RDF Schema) (W3C 2004) and OWL (Web Ontology Language) (W3C 2009) were able to deal with these issues quite effectively. The following are just a few examples of how these are dealt with in this case study.

In the AIA 201 Document clause 3.12.8, dealing with the "deviation existing between approved submittals and the Contract Documents", there is a default priority of the two in terms of the requirements to be complied with. For this situation, the type resource of `rdf:Seq`, one of the RDF Containers, is a suitable construct to express this kind of priority, since it represents a group of resources or literals in a certain order. This ordered group allows the user to determine the importance of any one of the resources with respect to the others, e.g. Contract Documents are more important than the Approved Submittals.

In RDFS, the property of `rdfs:subClassOf` can be used multiple times when defining a class. So, all the base classes introduced by `rdfs:subClassOf` will be `AND`ed together to create the new classes. Actually, this feature allows multi-inheritance, which exists a lot in the construction contract. For example, under the *Contractual Relationship* class *Promise* (see Figure 10-3), many concepts could be under both the authority and obligation classes. For example, the "Architect's enforcing the obligation against the Contractor", is the Architect's authority since there is no direct contractual relationship between the Architect and Contractor, while at the same time it is also an obligation of the Architect. The concepts with multi-inheritance are many under the *Contractual Relationship* class.

By using the set operators in OWL, new classes can be easily constructed by unions, intersections and complements of other existing classes. For example, the concept of the impact of a document on the "most recent schedules submitted to the Owner and Architect" in AIA A201 clause 3.10.3, can be defined by the intersection of the two sets of concepts: one set is all anonymous document classes with the property of `submittedDate`, and the other is the class schedule which represents all the schedules submitted and the selected value for this property is the date of the most recent schedule. This kind of class which needs a verbose description to be defined is quite common in the semantics of contract clauses.

Therefore, the ease of expressing the complicated concepts in contract clauses by the semantic modeling primitives of the ontology description language provides the potential and foundation for better domain knowledge representation. In addition, when performing the conceptualization development, being aware of the semantic modeling primitives' features will appreciably benefit the developer in identifying the underlying semantic structure in the textual material.

Applicability of the Taxonomy

Based on the conceptual model, the taxonomy developed extracts the relevant classes and organizes them systematically. It can serve as a structure for tracking concepts involved in contract management. The developed taxonomy, with more subclasses, instances and attributes added to enrich its semantic, can be used to develop an ontology which can provide more practical functionalities. For

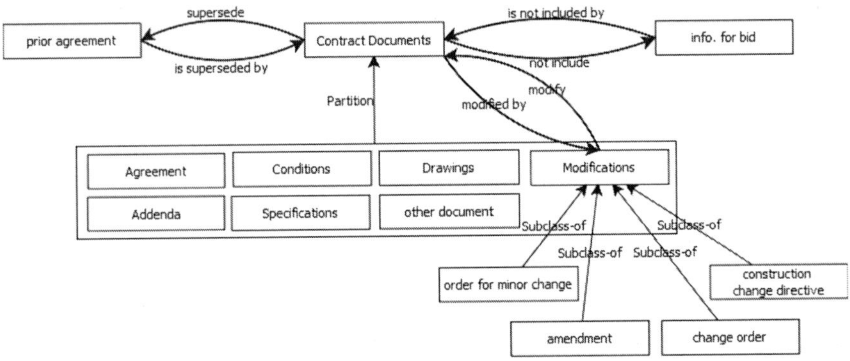

Figure 10-11. Excerpt of ontology for the concept of "Contract Documents"

example, as shown in Figure 10-11, the class "Contract Documents", based on its definition in AIA A201 clause 1.1.1, can be further developed into seven subclasses "Agreement", "Conditions", "Drawings", "Modifications", "Addenda", "Specifications" and "other document". All of them inherit the attributes of "Contract Document". Then if a specific instance is encountered, e.g. "Drawing A502, Exterior Doors and Windows Details", it will be identified as an instance of "Drawings" (Niu and Issa 2014). Because of the inheritance in the taxonomy, the subclass "Drawing" inherits the predefined attribute of "being legally binding" from its parent class "Contract Document". Thus, that piece of knowledge about the attribute of that specific instance can be obtained through this process. That explains how the taxonomy can make contributions to provide knowledge support in a particular application.

To implement the full effectiveness of ontology in a practical application for assisting contract management, other important technologies are also needed to allow the available knowledge to be properly utilized. As a particular application to achieve full effectiveness of ontology, focusing on the area of text processing to assist claim analysis, a rule-based NLP framework with ontology support has been proposed and a series of pilot studies has been conducted by Niu and Issa (2012 and 2014). Besides the ontological engineering aspect, from the review and adoption of impact factors-based paradigm in legal analysis, this framework integrates NLP technology and rule-based system to link certain language patterns to the entities of ontology, in order to interpret the semantics related to determining the impact factors' existences in a claim case.

CONCLUSION

The proposed taxonomy development methodology (mainly including building a conceptual model, common root classes from existing classification works, as well as the iterative approach and competency questions) is capable of guiding the taxonomy development process for the domain of construction contracts. The validity of the proposed methodology was verified by the taxonomy developed from the case study.

Furthermore, both the taxonomy development methodology and the taxonomy itself provide the bases for further ontology development in the domain of construction contract. The rest of the tasks in the ontology development for construction claims can proceed from this taxonomy, including defining the ad hoc relationships and attributes of concepts. Once the ontology is completed, legal analysis for construction contractual claims can be realized through a framework prototype. The developed ontology contains the conceptualized and formalized domain knowledge about construction contractual claims. It works as a knowledge base to provide original and explicit knowledge support for this system. Based on the knowledge representation in the ontology, an inference engine can do the reasoning work under certain circumstances. The implicit knowledge can be obtained as a result of the reasoning and added to the ontology as an extended and derived knowledge support system and software agents can be used to complete certain concrete and real legal analysis related jobs by utilizing the knowledge support derived from the interaction between the ontology and the inference engine. The interactions among these three basic modules constitute the fundamental mechanism of the framework prototype used for providing legal analysis services in the domain of construction contractual claims. In particularly, this framework prototype for legal analysis can be used to develop a series of applications for the practice of construction contracts and claims management. For example, a construction claim document production system can be developed by using the proposed ontology (Niu and Issa 2012). As noted earlier, by using ontology for representation of the domain knowledge, construction legal and claim analyses can be better realized than by using the pure rule-based expert systems developed in 1980s-1990s. Furthermore, based on the rule-based NLP framework supported by ontology, a application of the semantic interpretation of the impact factors in construction claim cases was also proposed and implemented (Niu and Issa 2014). In this application, a DSC Type I claim and its impact factors were chosen as the target to test the methodology. Through the combination of NLP, ontology as well as rule-based systems, the existence of impact factors in the text of a DSC Type I claim case history can be determined to improve the efficiency of legal analysis.

A taxonomy has been successfully developed as a result of the case study in order to validate the proposed taxonomy development methodology. But, this taxonomy has several limitations. First, it was only based on the first three out of the fifteen articles in the AIA A201-2007 Document. Although that is enough to validate the concept classification scheme of the developed taxonomy, the classification is to some extent incomplete. By further applying the proposed methodology and the experience of the case study, an enriched taxonomy based on the full content of the AIA A201-2007 Document can be developed in the future. Moreover, how to validate the developed taxonomy is another important consideration. Future efforts should focus on developing a practical and feasible validation method. Since taxonomy is a system for organizing concepts derived from professional expertise and insight, the validation method needs to also focus on the interaction with domain experts as well as take into consideration legal case precedents. In closing, the developed taxonomy in this case study should undergo the suggested improvements before it can be deployed to address actual construction contractual claim issues.

References

Adams, K. K. (1988). The development of an expert system for the analysis of construction contract claims. M.S. Thesis, University of Illinois at Urbana-Champaign, Urbana-Champaign, United States.

Alkass, S. T. and Harris, F. C. (1987). An expert system for claim analysis. *Proc., 3rd International Conference on Structure and Civil Engineering Computing*, Civil-Comp Press, 39–46.

Alkass, S., Mazerolle, M., and Harris, F. C. (1993). An integrated system to minimise the costs of analysing construction claims. *Computer System Engineering*, 4(2-3), 271–288.

Alkass, S., Mazerolle, M., Tribaldos, E., and Harris, F. (1995). Computer aided construction delay analysis and claims preparation. *Construction Management and Economics*, 13(4), 335–352.

Alshawi, M. A. and Hope, A. E. (1989). Expert systems and contractual disputes: Extension of time under JCT80. *Journal of Construction Management Economy*, 7.

Arditi, D. and Patel, B. K. (1989). Expert system for claim management in construction projects. *International Journal of Project Management*, 7(3), 141–146.

Arditi, D. and Pattanakitchamroon, T. (2008). Analysis Methods in Time-Based Claims. *Journal of Construction Engineering and Management, ASCE*, 134(4), 242–252.

American Institute of Architects. (2007a). General conditions of the contract for construction, AIA A201, Washington, D.C.

American Institute of Architects. (2007b). AIA Document A201 – 2007 Commentary. Washington, D.C.

American Institute of Architects. (2013). History of contract documents. http://www.aia.org/contractdocs/AIAS0766 71, September, 2013.

American Law Institute, A. L. (1981). Restatement (Second) of Contracts.

Aouad, G., Cooper, G., Brandon, P., Brown, F., Ford, S., Kirkham, J., Sarshar, M., and Young, B. (1994). *Integration of construction information (ICON) Final Report*, Department of Surveying & Information Technology Institute, University of Salford, Manchester, England.

Benjamin, P. C., Menzel, C. P., Mayer, R. J., Fillion, F., Futrell, M. T., deWitte, P. S., and Lingineni, M. (1994). *IDEF5 Method Report*, Knowledge Based Systems, Inc.

Berners-Lee, T., Hendler, J., and Lassila, O. (2001). *The Semantic Web*. Scientific American.

Björk, B. C. (1992), A Unified Approach for Modelling Construction Information. *Building and Environment*. Vol. 27, No. 2. pp 173–194.

Bob, W., Guus, S., Wouter, J., Anjo, A., and Harmelen, F. v. (1994). *Framework and Formalism for Expressing Ontologies*. University of Amsterdam.

Borst, W. (1997). Construction of Engineering Ontologies. PhD Dissertation, University of Twente, Enschede.

Bubbers, G. R. (1991). Computer assisted construction claim analysis. M.Sc. Thesis, University of New Brunswick, Fredericton and Saint John, NB, Canada.

Bubbers, G., and Christian, J. (1992). Hypertext and Claim Analysis. *Journal of Construction Engineering and Management*, 118(4), 716–730.

BuildingSMART. (2006). *IFC 2x3: Industry Foundation Classes*, ⟨http://www.buildingsmart-tech.org/ifc/IFC2x3/TC1/html/⟩, June 04, 2014.

Bullock, R. J. (1989). NOTICE: A Tutorial Expert System for Federal Government Notice Requirement Disputes. M.E. Thesis, Pennsylvania State University.

Chuan, C. and John, I. M. (2005). Taxonomy for International Construction Market Entry Mode Selection. *Construction Research Congress 2005*. American Society of Civil Engineers.

Cobb, J. E. and Diekmann, J. E. (1986). A claims analysis expert system. *Project Management Journal*, 17(2), 39–48.

Cooper, T. E. (1994). A knowledge-based construction claims advisor for the A.I.A. A201 General Conditions document. PhD Dissertation, Auburn University, Auburn, Alabama.

Cross, G. R. and deBessonet, C. G. (1985). Representation of legal knowledge for conceptual retrieval. *Information Processing & Management*, 21(1), 35–44.

Cunningham, H., et al. (2011). *Developing Language Processing Components with GATE Version 6 (A User Guide)*. University of Sheffield, Department of Computer Science, Sheffield, U.K.

Dagli, D. (1990). An expert system to provide pre-legal assistance in handling of potential claims under changes clause. MS Thesis, University of Colorado, Boulder, Colorado.

Diekmann, J. and Kim, M. P. (1992). Supercharge: expert system for analysis of changes claims. *Journal of Construction Engineering and Management*, 118(2), 399–410.

Diekmann, J. and Kraiem, Z. (1990). Uncertain Reasoning in Construction Legal Expert System. *Journal of Computing in Civil Engineering*, 4(1), 55–76.

Diekmann, J. and Kruppenbacher, T. (1984). Claims Analysis and Computer Reasoning. *Journal of Construction Engineering and Management*, 110(4), 391–408.

Diekmann, J. and Gjertsen, K. (1992). Site Event Advisor: Expert System for Contract Claims. *Journal of Computing in Civil Engineering*, 6(4), 472–479.

Diekmann, J. and M., Nelson (1985). Construction Claims: Frequency and Severity. *Journal of Construction Engineering and Management* 111(1): 74–81.

Lenat, D., Prakash, M., and Shepherd, M. (1985). CYC: Using Common Sense Knowledge to Overcome Brittleness and Knowledge Acquisition Bottlenecks. *The AI Magazine*, 6(4), 65.

El-Diraby, T. (2012). A Domain Ontology for Construction Knowledge (Dock 1.0). *Journal of Construction Engineering and Management*.

El-Diraby, T. A., Lima, C., and Feis, B. (2005). Domain Taxonomy for Construction Concepts: Toward a Formal Ontology for Construction Knowledge. *Journal of Computing in Civil Engineering*, 19, 394–406.

Emojong, A. (1987). An Application of Expert Systems to Legal Aspects of Construction. M. Sc.E. Thesis, University of New Brunswick.

Fischer, M. A., Betts, M., Hannus, M., Yamasaki, Y., and Laitinen, J. (1992). CIC Framework. *Workshop on Models for Computer Integrated Construction*, Finland: VTT, October.

Froese, T. M. (1992a). Integrated Computer-Aided Project Management Throught Standard Object-Oriented Models. Ph.D. thesis, Stanford University.

Froese, T. M. (1992b). GenCOM, A Domain Model for Project Management and Construction. *Workshop on Models for Computer Integrated Construction*, Finland: VTT, October.

Froese, T. (1996). Models of Construction Process Information. *Journal of Computing in Civil Engineering*, 10(3), 183–193. doi: 10.1061/(ASCE)0887-3801(1996)10:3(183).

Gjertsen, K. (1990). An expert system for claims classification. MS Thesis, University of Colorado, Boulder, Colorado.

Gomez-Perez, A., Corcho, O., and Fernandez-Lopez, M. (2004). Ontological Engineering: with examples from the areas of Knowledge Management, e-Commerce and the Semantic Web. First Edition (*Advanced Information and Knowledge Processing*), Springer.

Goodman, R. and Chinowsky, P. (2000). Taxonomy of Knowledge Requirements for Construction Executives. *Journal of Management in Engineering*, 16, 80–89.

Gomez-Perez, A., Corcho, O., and Fernandez-Lopez, M. (2004). Ontological Engineering: with examples from the areas of Knowledge Management, e-Commerce and the Semantic Web. First Edition (*Advanced Information and Knowledge Processing*), Springer.

González, A. J. and Dankel, D. D. (1993). *The engineering of knowledge-based systems: theory and practice*, Prentice Hall.

Gruber, T. R. (1993). A Translation Approach to Portable Ontology Specifications. *Knowledge Acquisition*, 5(2), 199–220.

Gruber, T. R. (1993). Towards Principles for the Design of Ontologies Used for Knowledge Sharing. In: *Guarino, N. and Poli, R., eds. Formal Ontology in Conceptual Analysis and Knowledge Representation, 1993*. Kluwer Academic Publishers.

Gruber, T. R. (1995). Toward Principles for the Design of Ontologies Used for Knowledge Sharing. *International Journal Human-Computer Studies*, 43, 907–928.

Grüninger, M., and Fox, M. (1995). Methodology for the Design and Evaluation of Ontologies. *IJCAI'95, Workshop on Basic Ontological Issues in Knowledge Sharing*, April 13, 1995.

Gruninger, M., Schlenoff, C., Knutilla, A., and Ray, S. (1997). Using process requirements as the basis for the creation and evaluation of process ontologies for enterprise modeling. *SIGGROUP Bull.*, 18, 52–55.

Guus, S., Bob, W., and Wouter, J. (1995). The KACTUS View on the 'O' Word. *International Joint Conference on Artificial Intelligence*.

Hackett, J., and Dancaster, C. (2000). *Construction claims: current practice and case management / by Jeremy Hackett*, London, LLP.

Hayes, P. J. (1985). The Second Naive Physics Manifesto. Formal Theories of the Common-Sense World, Ablex, Norwood.

Horridge, Matthew. (2011). *A Practical Guide To Building OWL Ontologies Using Protege 4 and CO-ODE Tools*, The University of Manchester, 1.3 ed.

Jackson, P. (1998). *Introduction to Expert Systems* (3rd Edition), Addison Wesley, Reading, MA, United States.

El-Adaway, Islan H., and Kandil, Amr A. (2010). Multi-agent System for Construction Dispute Resolution (MAS-COR). *Journal of Construction Engineering and Management, ASCE*, 136, No. 3, 303–315.

McCarthy, J. (1996). Some Expert System Need Common Sense. Computer Science Department, Stanford University.

Kim, M. P., Adams, K., and Florez, V. (1988). Differing Site Conditions Module for the Claims Guidance System: Initial Development of an Expert System. Construction Engineering Research Lab, US Army Corps of Engineers, Champaign IL.

Kraiem, Z. M. (1988). DISCON: An expert system for construction contract disputes. PhD Dissertation, University of Colorado, Boulder, Colorado.

Kraiem, Z. M., Dagli, D. P., and Diekmann, J. E. (1989). "DISCON: an expert system for the analysis of differing site conditions claims." *Knowledge-Based Systems*, 2(3), 158–164.

Kraiem, Z. M. and Diekmann, J. (1988). Representing Construction Contract Legal Knowledge. *Journal of Computing in Civil Engineering*, 2(2), 202–211.

Krippendorff, K. (2004). *Content Analysis: An Introduction to Its Methodology*, SAGE Publications.

Kululanga, G., et al. (2001). Construction Contractors' Claim Process Framework. *Journal of Construction Engineering and Management* 127(4): 309–314.

Lambe, P. (2006). Defining "Taxononmy", <http://www.greenchameleon.com/gc/blog_detail/defining_taxonomy/>, Straits Knowledge, June 07, 2014.

Lima, C., Diraby, T. E., Fies, B., Zarli, A., and Ferneley, E. (2003). The E-COGNOS Project: Current Status and Future Directions of an Ontology-Enabled IT Solution Infrastructure Supporting Knowledge Management in Construction. In: *Keith, R. M. and Paul, S.C., eds. Construction Research Congress 2003*. ASCE, 103.

Leondes, C. T. (2002). *Expert systems: the technology of knowledge management and decision making for the 21st century*, Academic Press.

Lopez, M., Perez, A., and Juristo, N. (1997). METHONTOLOGY: from Ontological Art towards Ontological Engineering. *Proc., Proceedings of the AAAI97 Spring Symposium*, 33–40.

Lopez, M. F., Gomez-Perez, A., Sierra, J. P., and Sierra, A. P. (1999). Building a chemical ontology using METHONTOLOGY and the Ontology Design Environment. *Intelligent Systems and their Applications, IEEE*, 14(1), 37–46.

Luiten, G., Froese, T., Björk, B.-C., Cooper, G., Junge, R., Karstila, K., and Oxman, R. (1993). An Information Reference Model for Architecture, Engineering, and Construction. *Paper presented at The First International Conference on the Management of Information Technology for Construction*, Singapore.

Luiten, G. (1994). "Computer aided design for construction in the building industry," PhD thesis, Fac. Of Civ. Engrg., Delft Univ. Of Technol., the Netherlands.

Mahfouz, T. and Kandil, A. (2009). Factors Affecting Litigation Prediction of Differing Site Conditions (DSC) Construction Cases: A Discrete Binary Choice Model. *2009 Construction Research Congress*, Seattle, Washington.

Mahfouz, T. and Kandil, A. (2010). Automated outcome prediction model for differing site conditions through support vector machines. *International Conference on Computing in Civil and Building Engineering*, Nottingham, UK.

Mahfouz, T. and Kandil, A. (2012). Litigation Outcome Prediction of Differing Site Condition Disputes through Machine Learning Models. *Journal of Computing in Civil Engineering*, 26(3), 298–308.

Neches, R., Fikes, R., Finin, T., Gruber, T. R., Patil, R., Senator, T., and Swartout, W. R. (1991). Enabling Technology for Knowledge Sharing. *AI Magazine*, 12(3), 16–36.

Niu, J. and Issa, R. R. (2012). Framework for Production of Ontology-based Construction Claim Documents. In: Issa, R. R. and Flood, I. (eds.) *ASCE International Conference on Computing in Civil Engineering*. Clearwater Beach, Florida: ASCE.

Niu, J. and Issa, R. R. (2013a). Conceptualizing Methodology for Building an Ontology for Construction Claim Knowledge. *ASCE International Conference on Computing in Civil Engineering*. Los Angeles, California: ASCE.

Niu, J. and Issa, R. R. (2013b). Taxonomy for the Domain Knowledge of Construction Contractual Relationships between the Owner and General Contractors. *Proceedings of the 20th International Workshop: Intelligent Computing in Engineer 2013*. Vienna, Austria: EG ICE.

Niu, J. and Issa, R. R. (2014). A Rule-based NLP Methodology for Semantic Interpretation of Impacting Factors in Construction Claim Cases. *ASCE International Conference on Computing in Civil Engineering and Building Engineering*. Orlando, Florida: ASCE.

Noy, N. F. and Mcguinness, D. L. (2002). *Ontology Development 101: A Guide to Creating Your First Ontology*. Stanford, CA: Knowledge System Laboratory.

Project Performance Corporation (2010). *Taxonomy Development Methodology*, <http://www.ppc.com/assets/pdf/white-papers/kmworld2010-day17-taxonomy-development-methdology.pdf>, June 07, 2014.

Riad, N., Arditi, D., and Mohammadi, J. (1991). A conceptual model for claim management in construction: An AI approach. *Computers & Structures*, 40(1), 67–74.

Rotar, M. A. (1990). Evaluation, testing and modification of an expert system for claims analysis under the changes clause. MS Thesis, University of Colorado, Boulder, Colorado.

Smith, G. R., Hanna, A. S., and Bullock, R. J. (1990). An Expert System Tutorial Approach for Evaluating Notice Requirement Disputes. *Engineering in our Environment: Proceedings Annual Conference of Canadian Society for Civil Engineering*, Hamilton, Ontario, Canada, 125–139.

Sowa, J. F. (1984). *Conceptual structures: information processing in mind and machine*, Addison-Wesley, Reading, MA, United States.

Staab, S., Studer, R., Schnurr, H. P., and Sure, Y. (2001). Knowledge processes and ontologies. *Intelligent Systems, IEEE*, 16(1), 26–34.

Studer, R., Benjamins, V. R., and Fensel, D. (1998). Knowledge engineering: Principles and methods. *Data & Knowledge Engineering*, 25(1-2), 161–197.

Suryanto, H., and Compton, P. (2000). Learning classification taxonomies from a classification knowledge based system, University of Karlsruhe.

Swartout, B., Ramesh, P., Knight, K., and Russ, T. (1997). Toward Distributed Use of Large-Scale Ontologies. *AAAI Symposium on Ontological Engineering*.

Thakker, D., Osman, T., and Lakin, P. (2009). *GATE JAPE Grammar Tutorial* (1.0 Ed.).

Thomas, J., Kelleher, J., and Smith, C. H. L. L. P. (2009). *Smith, Currie & Hancock's Common Sense Construction Law: A Practical Guide for the Construction Professional*, John Wiley & Sons.

Uschold, M. (1995). Towards a Methodology for Building Ontologies. *Workshop on Basic Ontological Issues in Knowledge Sharing IJCAI*. Montreal, Canada.

Uschold, M., and Grüninger, M. (1996). Ontologies: principles, methods, and applications. *Knowledge Engineering Review*, 11, 93–155.

Vidogah, W., and Ndekugri, I. (1998). A review of the role of information technology in construction claims management. *Computers in Industry*, 35(1), 77–85.

W3C. (2004). Resource Description Framework (RDF) model and syntax specification, a W3C Recommendation, 10 February 2004.

W3C. (2004). RDF Vocabulary Description Language 1.0: RDF Schema, a W3C Recommendation, February 10, 2004.

W3C. (2008). Frequently Asked Questions on W3C's Web Ontology Language (OWL), <http://www.w3.org/2003/08/owlfaq>, June 04, 2014.

W3C. (2009). OWL Web Ontology Language Overview, a W3C Recommendation, November 12, 2009.

Yu, L. (2011). *A Developer's Guide to the Semantic Web*, Springer-Verlag Berlin Heidelberg, Germany, Springer.

Index

Page numbers followed by *e*, *f*, and *t* indicate equations, figures, and tables, respectively.

actor and actor role taxonomy, 179–180, 180*f*
Actor Ontology, 172
AEC domain: classes and, 36; concepts and, 34; construction project contract management in, 218; instances and, 34; knowledge in, 107–108; legal document modelling and, 198–205; legislative data within, 194–195; pragmatic approach and, 28–29; regulatory document-user engagement within, 197–198; regulatory drafting and, 195–197; semantic interoperability and, 31; semi-automated ontology construction framework for legal ontologies in, 205–211; Serious Gaming and, 121–127
AI&CAR/TCA Reporting, 170–173, 171*f*, 175, 177, 183, 189
AKN (Akoma Ntoso), 202, 213
AKN XML, 202, 206–211
American Institute of Architects (AIA) Document A201-2007, 232–244. *See also* construction contract management case study
annotation, semantic, 137–139
Apache Jena Toolkit, 140
Architecture, Engineering, and Construction (AEC) domain. *See* AEC domain
Artificial Intelligence (AI): Deep Blue computer and, 101; ontology and, 98–99, 197, 220, 222
Asset Inventory and Condition Assessment Reporting or Tangible Capital Asset Reporting (AI&CAR/TCA). *See* AI&CAR/TCA Reporting
ATLAS.ti, 159, 163
axial coding, 159
axioms, 111–112

Borst, W., 222
Breuker, J., 199, 200, 203
BS6100, 29
Building Information Game (BIG), 129, 144, 146
Building Information Game Design (BIG Design), 129
Building Information Gaming (BIGing), 129
Building Information Modeling (BIM), Serious Gaming and, 122, 126, 128–130, 128*f*, 136, 139, 144
Building Project Model (BPM), 225

CAESAR Research Projects, 76
Cannan, J. P., 2
Casanovas, P., 197
CityGML, 165
classes, 35

CLIME Ontology, 202, 205
cognitive effect, positive, 30
Cohen's kappa measure, 5
coherentism, 106
collaborative tagging, 112
communication transaction-modality, 177–178, 177*f*
competency questions, in taxonomy development process, 231–232
concept modeling, 107
concepts, 34–35
conceptualization: explanation of, 150; informatics systems and, 109–110
conceptual model, of GUIs, 67–68, 68*f*, 68*t*
construction contract management: claim knowledge in expert systems and, 220–221; conclusions related to, 244–246; definition and methodology of ontology and, 222–223; domain knowledge and, 221; knowledge gaps and, 218; overview of, 219; solution exploration in, 218–219; taxonomy development methodology and, 226–232, 227*f*, 229*f*, 230*t*; taxonomy in ontology and, 223–226, 225*f*, 226*f*
construction contract management case study, 232; data preparation in, 233, 234*t*, 235; data processing in, 235–237, 236*f*; data selection, 233; results of, 238–244, 239*f*; tools used in, 237, 238*f*
construction industry, fragmentation in, 28
context modeling, 107
Contextual Passages, specifications for partitioning, 16–17
Controlled Natural Language (CNL) interface, 112
Core-Ontology classes, 40, 40*f*
Crowdsourcing, 98, 112
Crown Legislation Mark-up Language (CLML), 201–202
CYC, 101–102

data: explanation of, 102; unstructured, 110–111
data exchange, ontology and, 49
Davenport, T. H., 102
Declaration of Parliamentary Openness (OpeningParliament.org), 200–201
Deep Blue computer, 101
Dengel, A., 138
Description Logic (DL), 135
design, in sustainable construction, 74, 76–78, 77*f*, 78*f*
DILIGENT, 152, 153*t*, 154
Djaouti, D., 133–134
domain-knowledge-based IR systems, background of, 3–4
domain ontology. *See also* earthquake engineering base domain ontology: earthquake engineering, 6–15, 7*f*, 9*f*, 10*t*, 11*f*, 13*f*; general procedure to develop, 7–8, 7*f*; generating passages according to, 17, 18*f*; Wiki-Con, 34–35
domain taxonomies, 226
Domain Taxonomy Data Model, for healthcare facility management ontology, 62–63, 62*f*
domain theories: informatics ontologies as, 114; ontology as reasoning tool vs., 106
domain transaction-modality, 178–179, 178*f*
Dutch Legal eXtensible Mark-up Language (XML) Standard, 201

earthquake engineering base domain ontology: applied to NCREE reference collection to develop domain knowledge-bases IR system, 15–18, 18*f*; applying weighting rules for instance and

term filtering, 12–14, 13f; approach, 8–15, 10t; background, 6–8, 7f; discussions, 15; generating hierarchies and relationships, 10–12; glossary development, 9, 9f, 11f; NCREE reference collection and, 3–4, 4t; revising term relationships by domain experts, 14–15; revising upper-level concepts, 12
Earthquake Engineering Handbook (Chen and Scawthorn), 3, 8, 9, 12, 17
e-CKMI, 224
e-COGNOS, 6, 224, 226f, 230, 230t
El-Diraby, T. E., 180
electronic healthcare records (EHR). *See also* healthcare facility management ontology: ontology use and, 47–49
El-Gohary, N., 177
embedded annotation, 138, 139
end-user engagement: background of, 149–151; fieldwork-based ontology development methodology and, 155–160, 157f, 164–166; inner city subsurface utility reconstruction projects and, 160–164, 162t, 164t; ontology development and, 151–152, 153t, 154–155
Engineer-To-Order (ETO) products, 75
Enterprise approach, 152, 153t
ENTERPRISE Ontology, 222
e-society initiatives, 98
ethnographic research, 155–156
expert systems, 220–221, 220f
extended mind thesis (EMT): constructivism and, 116; knowledge and, 100; k-society and, 102; ontology style and, 112
extrinsic annotation, 138

facility management ontology. *See* healthcare facility management ontology
Failure Mode and Effects Analysis (FMEA), 53–55
Fault Tree Analysis (FTA), 53–55, 56t, 57f
FBX Software Development Kit, 130–131, 136
FBX technology, 130–131, 131f, 141
fieldwork-based ontology development, 155–160, 157f, 164–166
first-order logic, 109
Firth, J. R., 107
folksonomies, 112
formal concept analysis (FCA), 110, 113
foundationalism, 106
4D-CAD information system, 162, 165
4D implementation process, 162–165
FriendFeed, 109
Froese, T., 177, 179

Game Ontology Project, 135
gaming. *See* Serious Gaming
GATE (General Architecture for Text Engineering), 237, 238f
GIDS (Global Interlinked Data Store), 110
glossary development, 9, 11f
Goldcorp, 98
Gomez-Perez, 175, 185
Google page rank algorithms, 108–109
Graphical User Interface (GUI), 42, 43, 51, 65–68, 68f, 68t, 142–144
GRDDL (Gleaning Resource Descriptions from Dialects of Languages), 111
Gregory, J., 132
Gruber, T. R., 6, 101, 222
GUI mapping, 66

handbook-based ontology, 153t, 154
Harteveld, C., 125–127
Hartmann, T., 157, 157f

hCard, 109
healthcare facility management ontology: conceptual model organization and, 67–68, 68*f*, 68*t*, 69*f*; conceptual model system architecture and, 65–67, 65*f*, 66*f*; current literature and, 48–50; development of, 51–53, 52*f*; domain taxonomy organization and, 62–63, 62*f*; framework scope and development and, 60–61, 61*t*; GUI development and, 67; information analysis and, 53–55, 56*t*, 57–60, 57*f*, 58*t*, 59*f*, 60*t*; information exchanges and semantic reasoning and, 63–64, 64*f*, 65*t*; overview of, 47–48, 48*f*; research methodology and, 50–51; research results and, 70
hierarchy generation, 10–12
Hoekstra, R., 199, 200, 203–204
Huberman, A. M., 115
Huizinga, M., 122
Human Interface Devices (HID), 131

IBM, 101, 103
IDEF5 (Integrated DEFinition for Ontoloty Description Capture Method), 223
IFC, 29, 36–38, 136, 140*f*, 165, 226, 226*f*, 230, 230*t*. *See also* Industry Foundation Classes (IFCs)
IfcAnnotation, 139
IFC Java Toolkit, 136, 141
IfcPropertySingleValue, 139
impact assessment, 81, 84*f*, 84*t*
Industry Foundation Classes (IFCs), 29, 36–38, 74, 78–79, 83, 130
informatics systems: customization and, 109–110; function of, 103–104; introduction of ontology to, 98–99, 104–105, 105*f*
information analysis, healthcare facility management ontology and, 53–55, 56*t*, 57–60, 57*f*, 58*t*, 59*f*, 60*t*
information exchanges, for healthcare facility management ontology, 63–64, 64*f*
information retrieval (IR) systems, domain-knowledge-based, 3–4, 23
information taxonomy, 180–181, 180*f*, 181*f*
Information Technology (IT), 195, 197
Infrastructure and Construction Process Ontology, 172
infrastructure management. *See also* Transaction Domain Ontology (Trans_Dom_Onto); transaction formalization: background on, 170–171; protocol to formalize transactions in, 170–172, 171*f*
Infrastructure Product Ontology, 172, 173
inner city subsurface utility reconstruction projects, 160; domain description and, 161; process description and, 161–164, 162*t*, 164*f*
InnoCentive, 98
instances, 34–35
integrated project delivery (IPD), 74
intrinsic annotation, 138
inventory analysis, 79–81, 82*t*–83*t*, 83*f*
IRMA model, 224–225, 225*f*, 230, 230*t*
ISO Standard 10303, 115
Issa, R.R.A., 6
iterative development, in taxonomy development process, 231–232

Jaccard distance equation, 42*e*

KACTUS, 223
knowledge: contextual, 102–103; distributed, 102; encapsulation of, 107–108; explanation of, 102; social, 103–104

knowledge management, 104–105, 105*f*, 224
k-society (knowledge society): constructivism and, 116; extended mind thesis and, 102; function of, 98, 99; informatics systems and, 101, 104; innovation and, 104–105

LCA-related concepts. *See also* life cycle assessment (LCA): ontology development and, 84*f*, 79–81, 80*t*, 81*f*, 82*t*–83*t*, 83, 85*f*, 85*t*
LEED (Leadership in Energy and Environmental Design), 74, 75
legal mark-up languages, 201–202
legal ontologies, 202–205
Legislative Informatices, 194
Leonto, 205–212
life cycle assessment (LCA): function of, 75–76; integrated analysis and, 93; ontology development and, 84*f*, 79–81, 80*t*, 81*f*, 82*t*–83*t*, 83, 85, 85*f*, 85*t*
LKIF-Core (Legal Knowledge Interchange Format (LKIF)-Core), 203–205
LRI-Core, 203–205
Lynham, S. A., 114

Markov chains, 109
McGonigal, Jane, 121
message taxonomy, 179
metadata: ontology and, 108–109; semantic annotation and, 137–139
Methontology, 152, 153*t*
METHONTOLOGY, 223
Miles, M. B., 115
minimalism, 115
modal logic, 109
modeler input, WIKI and, 36–38, 37*f*
Mozilla Raindrop, 109
Mutis, I., 6

National Center for Research on Earthquake Engineering (NCREE), 3, 12, 14. *See also* NCREE reference collection
Natural Language Processing (NLP), 112
NCREE reference collection: background, 3–4, 4*t*; discussions, 5–6; document relevance assessments, 4–5; earthquake engineering base domain ontology applied to, 15–18, 21; information requests, 4–6, 5*t*
network theory, 112
nodes, 35
non-LCA-related concepts, ontology development and, 83, 85, 85*f*, 85*t*
Noy and McGuinness' methodology, 223
NVivo, 159

OAuth, 109
OKAPI BM25 model, 21
OmniClass, 29, 38
On-To-Knowledge methodolgy, 223
ontologies: concept modeling vs. context modeling and, 107; context-aware computing and, 49; customization and, 109–110; definitions of, 12–13, 99–101, 149–150, 175, 220–221, development of informatics, 114–116; as encapsulator of knowledge vs. checker of knowledge, 107–108; as enumeration of what is true, 108; integration and merger support and, 110–111; legal, 202–205; life cycle assessment, 75–76; metadata interaction and, 108–109; Protégé, 135–136, 186; as reasoning tool vs. domain theory, 106; structure of, 111–112; style of, 112–114; to support integration in sustainable construction, 74–75; taxonomy in, 223–226; term relationship

revisions and, 14–15; weighted rules and, 12–14
Ontology Annotation Tool (OAT) (GATE), 237, 238*f*
ontology architecture, transaction formalization and, 175–176, 176*f*
ontology development: background of, 78–79; end-user engagement in, 151–152, 153*t*, 154–155; end-user perspectives and, 151; fieldwork-based, 155–160, 157*f*, 164–166; for integrated analysis, 85–86, 87*f*, 88; k-society and, 104–105, 105*f*; LCA-related concepts and, *84f*, 79–81, 80*t*, 81*f*, 82*t*–83*t*, 83, 85*f*, 85*t*; non-LCA-related concepts and, 83, 85, 85*f*, 85*t*; outlook for, 97
ontology development methodology: end-user engagement and, 155–160, 157*f*, 164–166; semantic systems and, 114–116; transaction formalization and, 173–175, 174*f*
Ontology Engineering Methodology, 153*t*, 154
OntoPassage: development of, 17–18, 18*f*; purpose of, 23; search engine, 19–22, 20*f*, 22*t*
open, linked data (OLD), 196
OpenCalais, 112
open coding, 159
Open-electronic Data Interchange Transaction Ontology, 172–173
opinion dynamics analysis tools, 112
Osman, H. M., 182
overlap value, assessment, 5
OWL, 37–38, 135, 140, 141, 242, 243

Palmirani, M., 199
paper prototyping, 66
Particle Swarm Optimization, 109
passages, partitioning documents into, 16–17
POAF (Portable Ontology Aligned Fragments), 111
positive cognitive effect, 30
pragmatic approach: information sharing facilitation and, 28–29; semantic interoperability and, 44–45; for semantic reconciliation, 29–32, 31*f*; Wiki-Con as technical implementation of, 32–33
Prentsky, M. R., 123–124
prognostic normative reasoning (PNR), 111
Protégé ontology, 135–136, 186
Protégé-OWL, 135, 140, 141, 144
proto-type-oriented modeling approach, 113
Prusak, L., 102
Public Sector Accounting Board (PSAB), 170, 182

qualitative data analysis software, 159

RDF format, 111, 113, 114
regulatory drafting, overview of, 195–197
regulatory ontology construction: AEC regulatory document-used engagement and, 197–198; AEC regulatory drafting domain and, 195–197; background and conclusions related to, 194–195, 213–214; building regulations case example in, 205–211, 207*f*; discussion on, 212–213; legal document modelling and AEC domain and, 198–205
relevance theory, 30
Rezgui, Y., 74–75
RIF-BLD (Rule Interchange Format - Basic Logic Dialect), 204

Sartor, S., 197
search engines: for domain-specific documents, 2–3; general-purpose, 2; OntoPassage, 18–22, 20*f*, 22*t*
selective coding, 159

Semantically Interlinked Online Communities (SIOC), 114
semantic annotation, 137–139
semantic interoperability: explanation of, 30–31; pragmatic approach and, 44–45; research on, 29, 31
semantic reasoning, for healthcare facility management ontology, 63–64, 65*t*
semantic reconciliation: independent knowledge contexts and, 32; pragmatic approach for, 29–32, 31*f*; Wiki-Con and pragmatic approach for, 32–33
semantic systems: changes to shape ontology development and, 104–105, 105*f*; customization and, 109–110; evolving knowledge and, 98–99, 116–117; foundationalism and, 106; integration and merger support and, 110–111; interaction with metadata and, 108–109; ontology development methodology and, 114–116; path from data to information to knowledge and, 102–104; role of ontology and, 106–108; structure of ontology and, 111–112; style of ontology and, 112–114; Watson computer and, 99–102
SENSUS Ontology, 223
Serious Gaming: building information gaming and, 127–134, 130*f*, 131*f*, 133*f*, 133*t*; design model and, 144, 145*f*, 146; environment and, 139–144, 140*f*–144*f*; outlook for, 146, 146*f*; overview of, 122–125; semantic technologies and, 135–139, 136*f*–138*f*; use cases and, 125–127, 125*f*
Serious Human Rescue Game (SHRG) (Technische Universität Darmstadt), 127, 131, 134
Shields, P. M., 115
Simple Knowledge Organisation System (SKOS), 114
SimulElCon, 88–89, 89*f*, 90*t*, 91–92, 91*f*, 92*f*
SL (subsidiary legislation), 194, 196, 198, 205–211
social dynamic, 31
social media, 98
social networks, 98, 108, 109
software applications, ontology and integration of, 78–79
SPARQL, 114
special-purpose search engines, function of, 2–3
speech act theory, 30
Statistical Passages, specifications for partitioning, 16
sustainable construction: case study in, 88–89, 89*f*, 90*t*, 91–92, 91*f*, 92*f*; design in, 74, 76–78, 77*f*, 78*f*; explanation of, 73–74; LCA-related concepts and, 84*f*, 79–81, 80*t*, 81*f*, 82*t*–83*t*, 83, 85*f*, 85*t*; life cycle assessment and, 75–76; non-LCA-related concepts and, 83, 85, 85*f*, 85*t*; ontology development and, 78–81, 80*t*, 81*f*, 82*t*–85*t*, 83, 83*f*–87*f*, 85–86; ontology to support integration in, 74–75
SWEDER (Semantic Wrapping of Existing Data Sources with Embedded Rules), 110–111

tagging, 112, 138
Tajalli, H., 115
Tangible Capital Asset Ontology (TCA_Onto): application of, 184–185; architecture of, 175–176, 176*f*; background and conclusions related to, 171, 173, 181, 181*f*, 189–190; development of, 173–175, 174*f*; taxonomies and, 181–183, 181*f*, 182*f*
taxonomy development. *See also* construction contract management;

construction contract management case study: case study of, 232–233, 234*t*, 236*f*, 238*f*, 239*f*; discussion of case study in, 241–244, 244*f*; introduction to, 218–219; literature review and, 220–226, 220*f*, 222*f*, 225*f*, 226*f*; methods for, 226–232, 227*f*, 229*f*, 230*t*; in ontology, 223–226
TCA_Onto. *See* Tangible Capital Asset Ontology (TCA_Onto)
TCEI-Ontology: application of, 93; concepts in, 79; LCA and, 81, 83, 89, 91
technical documents: problems related to, 15–16; for specific domains, 2–3
Technische Universität Darmstadt, 127
temporal logic, 109
test-case validation, for healthcare facility management ontology, 67, 69–70
Text2Onto, 113
3-D modeling, Serious Gaming and, 129–132
time, cost and environmental impacts (TCEI). *See* TCEI-Ontology
TOVE (Toronto Virtual Enterprise), 152, 153*t*, 222
transaction attribute taxonomy, 180–181, 181*f*
Transaction Domain Extended Ontology, 175, 176*f*
Transaction Domain Kernel Ontology, 175, 176*f*
Transaction Domain Ontology (Trans_Dom_Onto): actor and actor role taxonomy and, 179–180, 180*f*; application of, 183–184; background and conclusions related to, 171, 176–177, 189–190; development methodology and, 173–175, 174*f*; information taxonomy and, 180–181, 180*f*, 181*f*; message taxonomy and, 179; transaction taxonomy and, 177–179, 177*f*, 178*f*
transaction formalization: background and conclusions related to, 170, 189–190; ontology application and, 183–185, 184*f*; ontology architecture and, 175–176, 176*f*; ontology development and, 172–173; ontology development methodology and, 173–175, 174*f*; ontology evaluation and, 185–189, 186*t*; protocol for, 170–172, 171*f*; tangible capital asset kernel ontology and, 181–183, 181*f*, 182*f*
transaction taxonomy, 177–179, 177*f*, 178*f*
Transaction Upper Ontology, 175, 176*f*
Trans_Dom_Onto). *See* Transaction Domain Ontology (Trans_Dom_Onto)
Triadic Game Design (TGD), 125–126
triangulation, 115
Trigger, 132–133, 133*f*

UML Use Cases, 53
Unified Modeling Language (UML), 55
Unreal Development Kit (UDK), 131
URIs, 113
Uschold, M., 222, 222*f*
use case development, 66
user query function, to communicate semantic content, 42–44, 43*f*

Watson computer, 101
Web Ontology Language (OWL), 135
Wiki-Con: associating semantic resource to core-ontology and, 38–41, 39*f*, 40*f*; communicating semantic content and user query function and, 42–44, 43*f*; concepts

and instances and, 34–35; core ontology approach and, 36; function of, 33–34; modeller input and, 36–38, 37*f*; relations and, 35–36; as technical implementation of pragmatic approach, 32–33; use and community of practice input and, 38

XFN, 109
XHTML data, translated into RDF format, 111

Zeb, J., 177, 179
Zhang, J., 180